W9-BHL-515

Technology Made EASY

Simple Ways to Use Your Computer, Phone, and Other Electronic Devices

Publisher's Note

This book is intended for general information only. It does not constitute medical, legal, or financial advice or practice. The editors of FC&A have taken careful measures to ensure the accuracy and usefulness of the information in this book. While every attempt has been made to assure accuracy, errors may occur. Some websites, addresses, and telephone numbers may have changed since printing. We cannot guarantee the safety or effectiveness of any advice or treatments mentioned. Readers are urged to consult with their professional financial advisors, lawyers, and health care professionals before making any changes.

Any health information in this book is for information only and is not intended to be a medical guide for self-treatment. It does not constitute medical advice and should not be construed as such or used in place of your doctor's medical advice. Readers are urged to consult with their health care professionals before undertaking therapies suggested by the information in this book, keeping in mind that errors in the text may occur as in all publications and that new findings may supersede older information.

The publisher and editors disclaim all liability (including any injuries, damages, or losses) resulting from the use of the information in this book.

The mind of sinful man is death,
but the mind controlled by the Spirit is life and peace.

Romans 8:6

FC&A Publishing®
103 Clover Green
Peachtree City, GA 30269

Produced by the staff of FC&A
ISBN 978-1-935574-44-6

Table of Contents

Apple OS X

Word & Excel

Apps
Online services that can change your life 261

Movies & TV

Music

The ever-changing tech world

Just when you think you're on top of the latest technology — BAM! — some new gizmo or upgrade or operating system comes out. That's when you start dreaming of simpler days, when winding the tape on your favorite cassette with your pinkie was your biggest tech challenge. But fear not — this book is here to help.

Technology connects you with family and friends, keeps your mind engaged and challenged, and offers a new and exciting creative outlet. Don't turn your back on these opportunities simply because you can't figure out the newest gadget.

Think of all the amazing things technology already lets you do. You can snap and send a picture of that broken doo-hickey on your dishwasher to your husband at the home center, then find the best deals for your prescriptions and order them online, and finally read your granddaughter a bedtime story via video chat — all in the same day. It's mind-boggling.

And what lies ahead reads more like science fiction than reality. Alarm clocks that go off early if there's roadway construction on your route to work. Medicine containers that alert you if you've forgotten to take your meds. A refrigerator that updates your grocery list by reading bar codes on food items as you remove them. Astonishing, but not far-fetched.

We live in truly amazing times, in part because we embrace change. Which is good, because technology changes as quickly as a Kansas tornado. Thankfully, with technology there's no destruction and no aftermath to clean up. It's just a matter of keeping up.

And people are certainly trying. Did you know more adults are actively using modern technology than not — regardless of age? That's right. Whether you are a Millennial, a Boomer, or of the G.I. generation, gee-whiz gadgets and the Internet are most likely part of your life.

By now, you probably know how to take a picture with your smartphone, order a book with your tablet, and create a to-do list on your computer. Good for you! Now let's take all that one step further. Turn the page and find hundreds of handy tricks for using your gadgets better, smarter, and easier.

The tips in this book are based on the following devices, operating systems, and programs. If you've upgraded operating systems or have different versions of these gadgets, don't worry — most tips will still apply. Some steps may not line up exactly, but they'll get you close enough to figure out what to do.

- iPhone 6 and iPad Air 2 using iOS 9.2
- Samsung Galaxy S6 phone and Galaxy Tab S tablet using Android 5.0 Lollipop
- Apple's El Capitan operating system
- Microsoft Windows 10
- Word and Excel in Office 2016
- Web browsers Safari, Chrome, Firefox, and Microsoft Edge

The digital revolution will be with us for a long time to come. Embrace the ever-changing tech world, and you'll go places you've never gone before. Strap in and enjoy the ride!

Compute like a pro

Slick tricks for PCs and Macs

Undo basic mistakes with one keystroke

You just finished writing a long email, but with an accidental swipe of the mouse, a big chunk of it disappeared. Uh-oh. Luckily you have an "undo" key on your computer, and it will correct all sorts of "oops" moments. Deleted a photo? Dragged something to the Trash or Recycle Bin you shouldn't have? Just press ⌘+z on a Mac or Ctrl+z on a Windows computer, and that oops moment can be your little secret.

Learn and create shortcuts for your keyboard

Keyboard shortcuts can be lifesavers, giving you faster and easier ways to get around on your computer. But how do you find out what they are? You can go to the Microsoft or Apple websites and print out lists of shortcuts, or just learn them by looking at the drop-down menus in each program you use.

Windows. The Windows logo key looks like this: ⊞. You'll see it used in several shortcuts, or you can use a combination of Ctrl, Alt, or Shift with other keys. A shortcut for closing windows, for example, is Ctrl+Shift+w.

Check out more shortcuts on page 84 in the Microsoft Windows 10 chapter.

Macs. Command, also called the "butterfly" key — ⌘ — is a symbol unique to Macs. Press it at the same time as other keys, and you unlock a world of shortcut possibilities. For example, the shortcut for closing windows is ⌘+w. Mac keyboards have a few other unique symbols too, so a cheat sheet might come in handy. Find one at *support.apple.com/en-us/HT201236*.

Customize Mac shortcuts. To create your own shortcuts or change the default shortcut settings on your Mac, go to > System Preferences > Keyboard > Shortcuts.

Apple also lets you create shortcuts for text you use often. For example, you can make the word "Love" pop up when you type the letter "L."

Go to > System Preferences > Keyboard > Text, and click the plus sign at the bottom to create as many as you like. Think "btw" for "by the way" and "ttyl" for "talk to you later." (See example on the next page.)

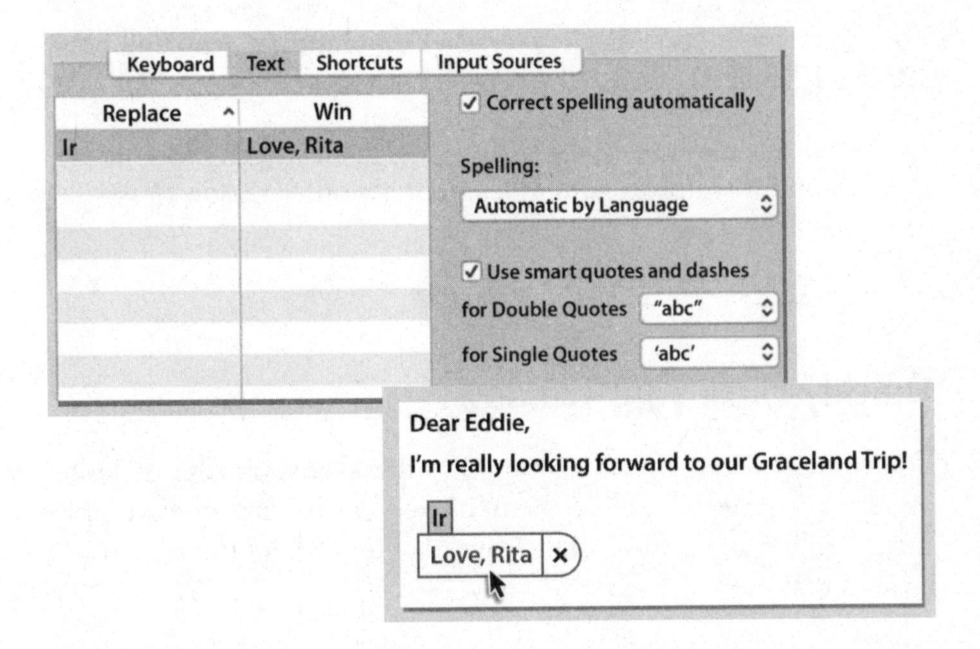

Force quit a frozen program

You pop over to your computer to quickly print out a coupon, but suddenly nothing happens. Is the program frozen? Sometimes it just takes a few seconds longer to run. But if you're sure you've reached the point of no return, it's probably time to force quit.

On a Windows computer. Press Ctrl+Shift+Esc. The Task Manager will open and show which programs are running, including those that aren't responding. Choose the problem program and click *End task* or *End now*. You can then reopen the program. (For more about "thawing" a frozen PC, see the tips on page 104.)

On a Mac. Press Option+⌘+Esc. A *Force Quit* box will pop up. Select the program and click *Force Quit*. If the program is on your

Dock, you can also right-click (or click and hold) its icon and press *Option*. Choose *Quit* from the drop-down menu.

Force quitting can cause you to lose unsaved changes, like edits to a photo or a word processing document. Another reason to save early and save often!

Quickly close your windows

Sometimes you end up with a ton of windows open on your desktop. When you're ready to close them, it's tedious to hunt down every little X in the top-left or -right corners. Good thing there's a way around that.

Windows. Press Alt+F4 as many times as you need to until everything is clear. To close out of not just the windows, but the whole app, press ⊞+q. Want to also lock your computer quickly when you walk away? Press ⊞+l.

Mac. Press ⌘+w to close windows. If you want to close out of apps, it's ⌘+q. You can quit all of your apps and lock your Mac by pressing ⌘+Shift+q.

Quit programs without clicking them

If you have lots of apps or programs open, there's a speedy way to close them from your PC's Taskbar or your Mac's Dock. Just right-click on the app's icon and you'll have the option to click

Close window on your PC or *Quit* on your Mac. If your mouse doesn't right-click, try Ctrl+click on either your PC or Mac.

Save your sight with these simple settings

Staring at a computer screen tires your eyes, so squinting is even worse. Is the text just too small? Or are you having trouble seeing the cursor? Customize settings and give your eyes a break.

Increase the size of your cursor. On your PC, go to ■ > Settings > Ease of Access > Mouse. Click to change your mouse pointer size (and even color).

On a Mac, go to > System Preferences > Accessibility > Display. Run out the sizing bar next to *Cursor size* until it's as big as you need. Also, check the box next to *Shake mouse pointer to locate.* To find the cursor again after your computer goes to sleep, just shake the mouse and the arrow will be huge for a few seconds to say, "Here I am!"

Zoom in for a closer look. On the PC, go to ■ > Settings > Ease of Access > Magnifier, and turn it on. Click the magnifying glass, and choose where and how much you want to magnify the screen.

On a Mac, go to > System Preferences > Accessibility > Zoom. Check *Use scroll gesture with modifier keys to zoom.* Now you can just press Control and scroll on your mouse to zoom — it works whether you have an actual scroll wheel or a Magic Mouse.

Scrolling up zooms in, scrolling down zooms out. This magnifies your entire computer screen whenever you want. You can also use the keyboard shortcuts listed under *Use keyboard shorts to zoom.*

Make everything bigger. Increase the size of everything on your PC's desktop by up to 175percent by going to ⊞ > Settings > System > Display > Customize your display. Under *Change the size of text, apps, and other items: 100% (Recommended)*, drag the bar from the left all the way to the right.

On a Mac, go to > System Preferences > Displays. The resolution is probably set to *Default for display*, but you can click *Scaled* and choose a different resolution to increase the size of icons, text, and programs.

Increase text size in a program. In many Apple apps and programs, you can increase text size by pressing ⌘ and the plus key or decrease it by pressing ⌘ and minus. Or check under the app's Preferences. Only need to make things bigger when you're surfing the Web? See the tip on page 218.

Right-click your mouse for more menus

Even if you're computer savvy, you might not know the shortcuts you can take by right-clicking your mouse.

Windows mice usually have left and right buttons. Plus, laptop trackpads sometimes have a button specifically for right-clicking or you might have to double-tap for the equivalent of a right-click.

Apple mice come with their right-click button disabled. To get it up and running, go to > System Preferences > Mouse > Secondary Click > Click on right side (or *Click on left side* if you prefer).

If you're not a fan of the right-click, press Control while you click as an alternative. On a trackpad, you might have to double-tap.

Depending on the program, right-clicking can bring up menus that let you:

- create a new folder on your desktop or within any other folder.
- move a file to the Trash or Recycle Bin.
- move to another application.
- look up the meaning of a word.
- reply to or forward an email.

New way to cut and paste — no glue required

Cut and paste. Copy and paste. We perform these basic tasks more times a day than we can count. There are several different ways you can do it, but many people highlight text, then use ⌘+c and ⌘+v on a Mac, or Ctrl+c and Ctrl+v on a Windows computer. But here are a couple of tricks you might not know.

Copy and move. Try highlighting the text as usual, then clicking it while holding down Option on a Mac or Ctrl on a PC. Drag the text to where you want it (you'll see a plus sign), and then let go of your mouse.

Cut and move. On a Mac, highlight the text, then hold down ⌘ and drag it. This will cut the text out of the document, and paste it wherever you release the mouse.

On a PC, highlight the text then right-click. Pick *Cut* from the drop-down menu. Move your pointer to where you want to move your text. Right-click and pick from one of three *Paste Options.*

These new moves might become your favorite ways to cut, copy, and paste while on your computer.

Find out what those 3 little dots (...) mean (if you like options)

Sometimes it's easy to miss something that's right under your nose. Or in this case, on your drop-down menu. When you go to a drop-down menu in a program and click an option such as *New* or *Open*, sometimes you see three dots (..., known as an ellipsis) next to the word. The ellipsis is a clue that the program has additional options. For example, if you're on a Mac using Preview and go to File > Open..., you'll get a dialog box, often a folder or location you last used in Preview.

You might also see a small right arrow or triangle to the left of some items on drop-down menus. That symbol means there's a submenu to click for more options. Going back to the Mac Preview example, there's a submenu under *Open Recent*, so you can conveniently choose from your most recent files.

Change your desktop image in seconds

Tired of looking at the same old image on your computer screen? Here's an easy way to turn your favorite picture into your computer's "wallpaper."

On a Mac. Go to > System Preferences > Desktop & Screen Saver. Choose from any picture in your Photos or Pictures folder, or click the plus sign at the bottom left to search a different folder.

Then choose how you want it to appear on the screen from the drop-down menu in the middle of your *Desktop & Screensaver* window. Choose *Center* to have a smaller photo sit in the middle of your screen framed in a colored background, or *Tile* to have repeats of the same image all over your screen.

If the photo is bigger, you can choose *Fill Screen, Fit to Screen*, or *Stretch to Fill Screen*. Your image could end up distorted depending on its size as well as the size of your display and its resolution, though, so going for the biggest option is a good rule of thumb.

On a Windows computer. Go to ⊞ > Settings > Personalization > Background. Choose *Picture, Solid color*, or *Slideshow*. Or click *Browse* to search in your Saved Pictures or Camera Roll folders, like if you want to use a photo taken with your synced smartphone, for example. Or right-click on photos saved to your computer and choose *Set as Background* from the drop-down menu.

Check out your pre-installed programs (and never buy software again)

If you just got a new computer, or even if it's just new-ish, you might not be taking advantage of its full potential. Before you spend money on programs, think about how you're going to use your computer. To check email? Surf the Internet? Listen to music? Don't shell out another dime — you can do it all with your computer's pre-installed programs.

On a Windows computer. The newest apps in Windows are all the rage. Check out these handy programs for PC lovers.

- Use One Note if you need to jot something down or create a checklist.
- Merge all of your email accounts on Mail to see all your emails in one central location.
- Organize your tunes in Music and pictures in Photos.
- Surf the Web with Microsoft Edge, the new default browser.
- Check out this week's forecast on Weather and the latest scores on Sports.
- Stay up to date with financial news on Money, or check out the headlines with News.
- Never get lost again with Maps powered by Bing.
- Manage all of your email, calendar, contacts, and tasks via Outlook.
- Need help with anything? Ask Cortana, Microsoft's virtual assistant.

On a Mac. You'll find a chapter full of tips in *Apple OS X: unleash the magic in your Mac*. But here's a quick rundown of Apple's built-in programs.

- Compose documents in Pages or jot down things in Notes.
- Set reminders in, handily enough, Reminders.
- Use Numbers for spreadsheets.
- Combine your email accounts into Mail.
- Go on an Internet Safari with the aptly named Web browser.

- Rock out to music in iTunes or make your own with GarageBand.
- Make and view movies in iMovie, and keep track of your pictures in Photos.
- Want to chat with relatives online? Try Messages and FaceTime.
- Use Maps to plot your way around.

Once you explore all of your built-in options, you might be surprised at how much you can already do.

Drag and drop files — when it copies and when it moves

So you just dragged an icon from one window to another. Did you move the file, or make a copy of it? Here's an easy way to tell the difference.

On a Mac. When a green plus sign (+) appears on the icon, that tells you you're copying the file. You'll see it when you press Option as you drag and release a file. If you don't see the plus sign, you're simply moving the file, not copying it.

On a PC. Right-click while dragging a file. A pop-up menu will ask if you want to *Move here* or *Copy here*. Or you can use two of your keys. Press Ctrl while dragging a file to copy it, or press Shift while dragging a file to move it.

If while moving a file you realize you meant to copy it instead, press Esc while you're still holding down the mouse or choose Cancel from the Windows pop-up menu.

Choose only the files you want to move

Sometimes you just want to pluck one file from a folder and move it. Or maybe you want to move everything but a file or two. Take advantage of your Mac's ⌘ button or your PC's Ctrl key to move only what needs to go.

Select everything in a list by pressing ⌘+a or Ctrl+a.

Select all but a few items by selecting all files first, then hovering your pointer over the file you don't want to move, and pressing click+⌘ or click+Ctrl.

Select consecutive items by clicking the first file you want to move, then pressing Shift while clicking the last file you want to move. You'll select the first file, the last file, and everything in between. If you change your mind about some of the files, press ⌘ or Ctrl while clicking them.

Windows has another technique for selecting consecutive items. Just hold down the mouse button while clicking slightly above the first, then drag the mouse down to point to the last file in the series. Let go, and everything caught in the "lasso" is selected. Get along, little files.

The secret to decluttering your computer

If you want to clean out the files on your computer, sort them first so you don't accidentally get rid of anything important. You can sort them by attributes like size or date added.

On a Mac, go to View > Arrange by. On a PC, right-click inside the folder, go to *Sort By*, and use the pop-up menu.

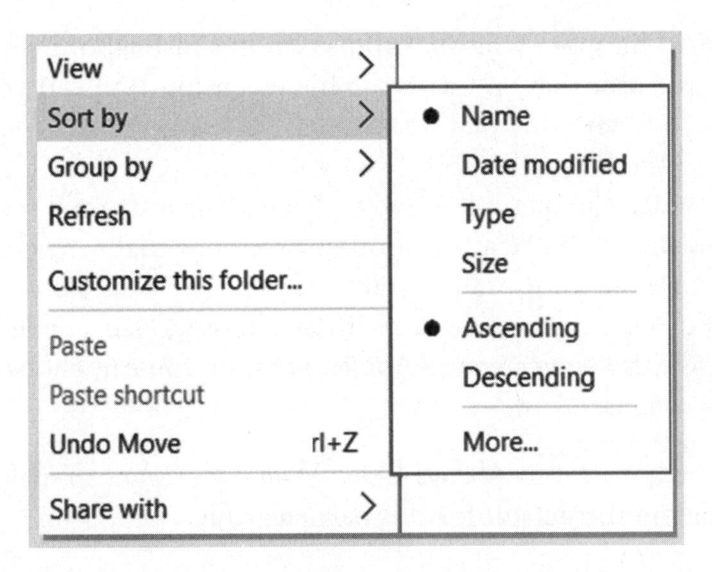

If you want a certain file or folder to always come first without giving it a name beginning with "A," name it a number instead or put a number in front of the name. Your computer will always sort files by numerical order before it sorts by alphabetical order.

On a Mac, click on the file name once, then move the mouse away quickly. Edit the name. On a Windows computer, click inside the name field and press F2.

Open files with your favorite programs

There's a default application or program for just about every kind of file. Want to listen to music on your Mac? Go to iTunes. Need to crunch numbers on a PC? Open Excel. But there are times when you probably want to change the default program, the one your computer always uses to open a specific type of file.

On a Mac. Suppose you want to use Photos instead of Preview to open pictures that end with the file extension PNG. If you just want to do it with one picture, you have a few options. Drag the picture to the Photos icon if it's in your Dock. Or, right-click or Control+click the picture. Choose *Open With* and you'll see the default program above a list of other programs that can open the file. Choose one of those programs, *Other...*, or *App Store...* if you want to get a new program. If you choose *Other...*, you'll get a pop up with *Recommended Applications*, or you can choose *All Applications.*

Check the box next to *Always Open With* if you want to make the new program the default for this particular file.

To change the default for all files of that type, right-click a file and go down to *Get Info*, or click the file once and press ⌘+i. Peruse the options under *Open with.* After making your selection, click *Change All...* under *Use this application to open all documents like this one.* For example, you can make all PNG files open in Photos instead of Preview by going through these steps.

On a PC. If you want to open one file with a different program, right-click on the file and choose *Properties.* Under the *General* tab, click the *Change...* button. Choose a program from the list and press *OK.*

To change the default for all of these types of files, right-click the file and select *Open with* to find another program. Don't see one you want? Go to *Choose another app* or *Search the store.* Once you find the app you want, click *OK.*

You can also set defaults by going to ⊞ > Settings > System > Default Apps. Each type of file is listed on the right. Click on a program's name to change the default.

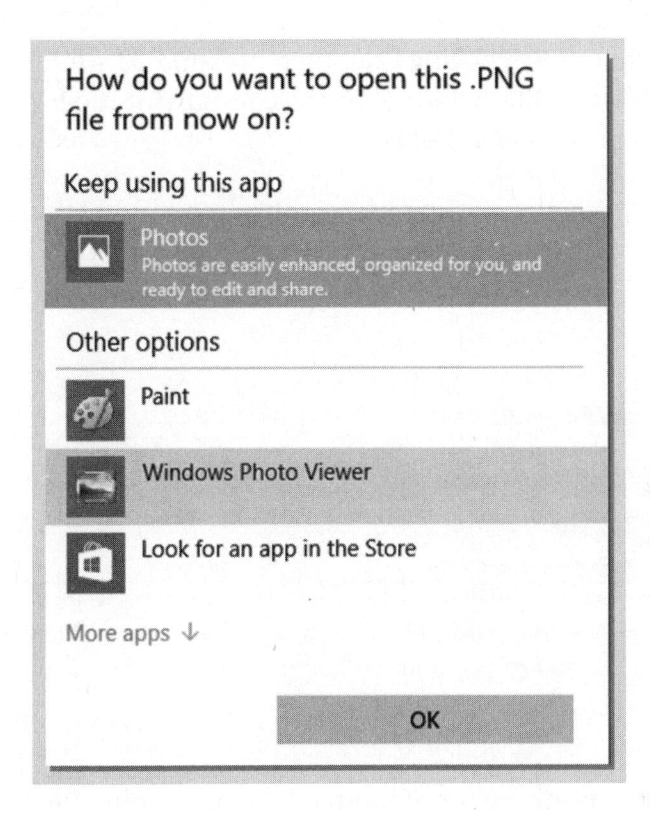

Rename lots of files all at once

You take tons of photos, but when you upload them to your computer, each name is just a string of characters. Other than the date, there's no way to tell your vacation photos from any other photos. Instead of renaming each one individually, Macs and PCs can rename photos in batches.

On a Mac. Open the folder containing the files you want to rename and select them. Choose multiple files by clicking the first one in the series and holding Shift while you click the last one. Or pick and choose by clicking one, then pressing ⌘ while you click the rest. Now right-click or Control-click any of the files and choose *Rename (number) of items.*

You can replace the existing name entirely, add something before or after the name, or give them a custom name followed by a number in sequence, like "beach2015_1.jpg," "beach2015_2.jpg," and so on.

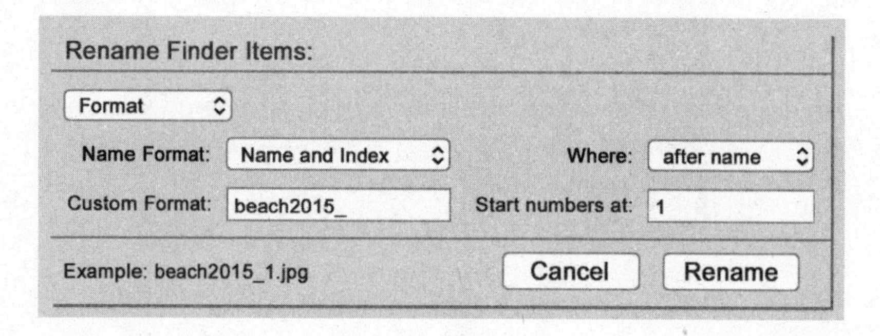

On a PC. Options are more limited when renaming files on a PC, but you can at least get some unique names. Select all of the images you want, then press F2. Rename one file and press *Enter*. The rest of the files will get the same name. If they're the same type of file, they'll be named in numerical sequence starting with 1.

Easily find those 'lost' files

It happens to the best of us. Download or save a new file and it seems to "disappear" into the depths of your hard drive. If you don't immediately see it, don't worry, it's still there. Try these tricks for finding that file that's lost somewhere in your computer.

On a Mac. Search for lost files by pressing ⌘ and the spacebar, or clicking the magnifying glass on the top-right corner of your screen. When the Spotlight search box pops up, type in all or part of the file's name.

Spotlight Search

Your computer will pull up its *Top Hits* along with other possible matches. Click *Show all in Finder* at the bottom of the search results window to choose whether to look other places, such as shared drives.

On a PC. Press ⊞+e to open a Desktop window. Look through the folders. No luck? Click ⊞, type the name in the search box, and press Enter.

But what if you can't remember exactly what you named it? Here are more ways to find lost files.

Limit your search. Both Macs and PCs search all of the folders on your computer (and elsewhere, if you want), but you can also limit the search. On a Mac, go to System Preferences > Spotlight and uncheck any program you don't want to include. On a PC, choose *My Stuff* or *Web* at the bottom of the search window. Narrow it down by using the drop-down menu next to *Show*.

Use quotation marks. If your file name contains more than one word, use quotes around it (like "dog howl"). You can also use NOT and OR (in all caps) if you want to include or exclude a word.

Put your Dock or Taskbar in its place

Your computer's Dock (Mac) or Taskbar (PC) is that bar that runs across the bottom or side of your desktop, storing the icons for the programs you use most often. Here's how you can move your Dock or Taskbar, and make it visible or invisible.

Taskbar. The Windows Taskbar has an auto-hide feature, meaning it won't appear until you hover your mouse pointer over it. If this bugs you, turn it off. Right-click somewhere on the bar and then click *Properties.* Click the box next to *Auto-hide the taskbar* and it'll stay visible. If it's on the side and you want it across the bottom, grab it with your mouse and drag it down. Keep it there by right-clicking and choosing *Lock the taskbar.*

Dock. To keep your Dock visible at all times, right-click or Control-click it and choose *Turn Hiding Off.* Make the icons pop up larger when you scroll across them by choosing *Turn Magnification On.* To change your Dock's placement, go to *Position on Screen* and check *Left, Bottom*, or *Right.* Choose *Dock Preferences* to experiment with its size.

Find apps easily by pinning them to your Dock/Taskbar

Tired of digging around in Finder (or going to your Start Menu, if you have a Windows computer) whenever you need a program? Your only option used to be placing shortcut icons to various programs on your desktop. Now there's a new shortcut in town, and it's bound to become your favorite way to find your apps. It's called "pinning" — placing an icon representing a one-click shortcut in a special place you can access at any time.

On a Windows computer. Right-click your favorite app and select *Pin to taskbar* in the pop-up menu. Now when you move your cursor to the Taskbar, you'll see a bar pop up with an icon for that program. It's a great way to easily access your most frequently used programs and apps. If you want to remove one, just right-click again and choose *Unpin from taskbar*.

Once you have several there, pretend they're numbered starting with the icon furthest to the left (or at the top, if your Taskbar is vertical), and press ⊞+1 to open the first app, ⊞+2 to open the second app, and so on.

On a Mac. To pin an app, click the Launchpad rocket ship icon in the Dock, find the app, and drag it into the Dock, or click and drag it from a folder. To remove, right-click it and choose Options > Remove from Dock. You can also drag the icon off the Dock and wait a few seconds. The word *Remove* will show up and the icon will disappear. Don't worry, the program lives on.

Pin programs, folders, files, and more for easy access

Remember when you were limited to creating shortcuts to programs only on your desktop? Now there are many different places to pin programs, apps, folders, and files. Here are a couple of locales where you can quickly access what you need, when you need it.

On a Mac. Open your Finder window. See the row of icons across the top?

That's called your Toolbar, and you can add just about anything you want there. Maybe, for instance, you'd prefer to have your Mail program on the Toolbar. Just hold ⌘ as you drag the Mail icon from your Dock to anywhere on the Toolbar until you see a green plus sign. Release your mouse, and the icon will be copied into the Toolbar. You can do this with a document you refer to repeatedly, too.

To take something out, drag it away while holding ⌘ and poof, it's gone.

On a Windows computer. In Windows 10, there's a whole lotta pinning going on. You can pin apps and programs to your Start Menu by right-clicking them and choosing *Pin to Start.* Pin frequently accessed folders to the *Quick access* area (what used to be called Favorites).

Go to File Explorer > Quick access. If a file or folder doesn't have a "push pin" next to it, right-click and choose *Pin to Quick access.*

Need to switch things around? Another right-click lets you *Unpin from Quick access.*

Flip from photo to photo with a simple keystroke

Say you've got a dozen photos open on your computer. Here's an easy way to flip from one to another without your mouse. On a Mac, hold ⌘ and then press the tilde key (~). It's near the top-left corner on your keyboard.

On a Windows computer, hold Ctrl and press Tab. This neat trick might work with other programs, too, like a window with multiple documents open.

Quickly flip back and forth between multiple programs

Do you like multitasking with many programs open at the same time? Hold down ⌘ and press Tab to quickly go through them on a Mac. A small window will pop open that looks an awful lot like your Dock. But take a closer look and you'll see it's only showing the icons of each open program. Apple calls this a heads-up display.

Keep clicking forward through the icons or go backwards by holding Shift+⌘+Tab. When you've highlighted the one you want, release ⌘. Your program will open and come to the forefront of your screen.

To switch between two programs, like a Web browser and an email program, briefly press ⌘+Tab to go to your most recent program. Press it again to go to the one you started in. Keep pressing them to "bounce" back and forth between the two programs.

In Windows you can do the same, but the screen display looks different. Windows calls it a map.

Click Alt+Tab and you will bounce from one open window to the next. Say, for instance, you want to flip from your calculator to a spreadsheet. This is a handy way to do that. Then release the Alt button and the program you landed on will pop open, front and center.

See your programs side by side in a 'snap'

Even if you already know how to flip back and forth between different programs or apps, getting two apps to run side by side on your screen can really boost your productivity. The latest versions of both Mac and Windows operating systems offer the ability to "snap" apps that will make the neatest neat freak deliriously happy.

On a Mac. Click the green button in the top-left corner of the app's Title bar. To snap it to one side of the screen, hold the green button and move your mouse left or right. You'll see a rectangular shadow appear on the screen. When you let go of the mouse, the window will snap into place. Grab another open window and drop it onto the other side of the screen. You'll notice both programs are in full-screen mode; it's the only way to use Split View on a Mac, so it won't work with programs that can't run full screen.

Split View not working? Go to > System Preferences > Mission Control, and check the box next to *Displays have separate Spaces.*

In Windows. Click and hold the top edge of a window (the Title bar), and drag it to the left side of your screen. A rectangular outline will appear. Release the mouse and drop it in place. Grab another window by its Title bar, then drag it to the right side and drop it.

Ditch your mouse altogether for snapping and use your keyboard. Send a window to the left by pressing ⊞+←, to the right by pressing ⊞+→, and full-screen by pressing ⊞+↑ until it hits the top.

Arranging windows like this lets you cut, copy, or paste between files. It's also perfect when you're moving files from one folder to another. Just snap your folders into side-by-side windows.

4 windows, 1 screen — a fun way to boost productivity

If you like the idea of side-by-side windows, try having three or four tiled across your screen at once. Windows makes it easy with Snap Assist.

In Windows, snap an open window to one side (either by dragging its Title bar to the left or right, or using ⊞+← or ⊞+→. Snap

Assist pops up and shows you small icons or thumbnails of the other windows available. Then you can choose which window to snap next.

Vertical snap. Choose one window and press ⊞+↑ or ⊞+↓. Then snap the other window either up or down the same way. To maximize a window, keep pressing ⊞+↑ and minimize by pressing ⊞+↓.

2 x 2 snap. You can also snap windows to create a 2 x 2 grid on your desktop. Grab the window with your mouse, then drag and drop it to one corner of your screen. Snap Assist comes up to help you fill the rest of the screen. Keep going until all four quadrants are filled.

Of course there are keyboard shortcuts. Press ⊞+↑ to snap a window up, then ⊞+→ to put it in the top-right corner, for example. Use the arrow keys with ⊞ to move windows around in the quadrants.

Triple snap. Want three windows open at the same time? Just snap two of the windows as described, and bring the third one either up or down so it's taking up the top or bottom half of the screen. Or make one tall window by moving a third to the left or right.

Too much? You don't have to fill up every empty space on the screen. If you leave a quadrant or half of the screen open, Snap Assist gives you the option to keep going, but you don't have to use it.

See what's on your desktop with one keystroke

Suppose you need to remind yourself of the name of a file you just saved to your desktop, or you're waiting to see if your download has come in, but you can't see the file because you have too many windows cluttering your computer screen. Learn this quick move to make those windows quickly jump out of your way.

PC users. Press ⊞+d to hide everything and reveal the desktop underneath. Click ⊞+d to reopen everything.

Here's another neat trick if you want to hide all windows but the one you're working on. Click and hold the top edge (Title bar) of the window and then shake it fast with your mouse, as if you're shaking out a towel. This simply shrinks the other windows down to tiny icons on your Taskbar. Get them back by shaking the open window's Title bar again or clicking their icons on the Taskbar.

Mac users. On Macs, it all depends on your keyboard. Press ⌘ +F3 or Fn+F11 to see what happens. On an older keyboard, you might be able to just use F3.

Use one printer for all of your devices

In many offices, several people share a single printer because they're part of a network. You can do the same thing at home and keep three devices from needing three different printers. To connect them all to the same printer, you will need a Wi-Fi-enabled printer and a Wi-Fi network set up at home.

Some Wi-Fi printers show up by themselves on other computers using the same network, so all you have to do is choose it after clicking *Print.* If you plug your printer into your computer using a USB port, you can still share it with other computers as long as it's on a network and has Wi-Fi capabilities.

On a Mac, go to > System Preferences > Printers & Scanners. Find the printer icon and check *Share this printer on the network.* Click Sharing Preferences > Users, and choose who (or what device) can print. Other users should be listed by whatever their computer or device is named.

To use a printer already connected to another computer, get your document ready to print. When the print dialog box opens, look for printer options in the drop-down menu.

On a Windows computer, right-click , then click *Control Panel.* Under *Network and Internet*, click View network status and tasks > Change advanced sharing settings. There you can turn printer sharing on or off.

To add a printer connected to a different computer, click > Settings > Devices > Printers & Scanners, and then click the plus sign next to *Add a printer or scanner.* The printer should show up.

Don't be greedy — simple ways to share your files

Sometimes you write for yourself, and sometimes you want to give another person a piece of your mind. In the form of a file, of course.

Microsoft and Apple understand you don't exist in a vacuum, so both offer a super simple way to share documents on your computer with anyone.

On a Mac, just look for the Share icon. It's a square with an arrow pointing up, and it'll show up in Pages, Numbers, Photos, and other Apple programs. When you click it, there's a drop-down menu with sharing options.

If you use iCloud, you can click *Share Link via iCloud* and share the file in AirDrop, Facebook, LinkedIn, Mail, Messages, or Twitter. You can also go to *Send a Copy* to share with Messages, Notes, AirDrop, or Mail. The program you choose will open with the file attached.

In Acrobat, a drop-down menu in the Print dialog box lets you quickly email a PDF.

On a Windows computer, click *File* while in the document, then choose *Share*. Pick a sharing option like Email > Send as Attachment, and your email program will open with the document attached. This will work in Outlook, Power Point, Word, and Excel.

Speed up your printing process

You're ready to print out tickets to a concert, but first you have to remember where you left the file on your computer. Then you double-click to open the file in a program and choose Print from the drop-down menu, or press ⌘+p. Usually you have to click Print again after making sure the settings are just right. It seems like it takes forever, but you can actually print without opening a program at all.

On a Windows computer. Right-click on the unopened document's icon and choose *Print*.

On a Mac. Go to > System Preferences > Printers & Scanners. Click your Printer icon and drag it to your Dock. The next time you want to print something, just drag it to the Dock and drop it on the Printer icon. It'll immediately start printing.

Permanently purge your print job

You change your mind all of the time. But when it happens after you've sent a 30-page manual to your printer, what do you do? You can run to the printer and hit the Off button. It'll stop, and you'll forget about it. Problem solved. Except the next time you — or anybody else — turns on the printer, it'll pick right up where it left off. To make it go away, you have to clear the print job entirely from the printer's memory.

On a Windows computer. Go to the Taskbar and right-click the printer icon (on the far right while the printer is operating). Click your printer's name, and a window opens with all the print jobs currently sent to that printer. Right-click the item you want to stop, and choose *Cancel* from the drop-down menu. You'll be asked if you're sure you want to cancel. Click *Yes*.

On a Mac. When you're in the middle of printing, the printer icon should be in your Dock. Click it and a printer dialog box will open. Click the "x" next to the print job you want to cancel.

If your document hasn't started printing yet or you don't see the printer icon in your Dock, go to > System Preferences > Printers & Scanners. Find the document in the print queue and click the "x."

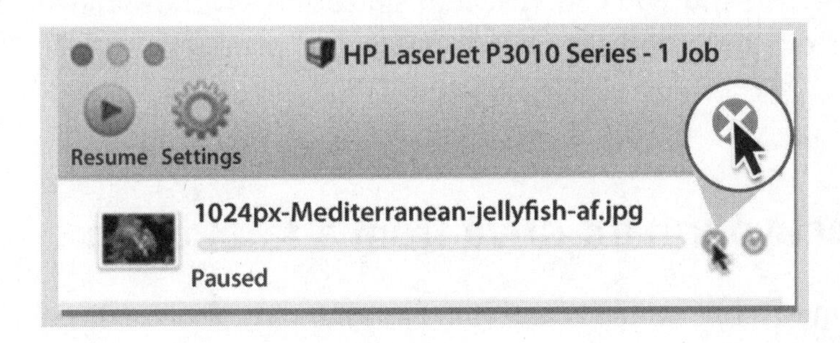

Save your files on your own personal server

It's easy to fill up the hard drive on your computer and run out of space. Turns out, it's also easy to protect all the important data on your computer and clear out a lot of room — but still keep your files, photos, and more.

Enter cloud drives. Instead of being kept somewhere tangible, you pay a fee for a company to keep your files on their servers so you can access them via the Internet. As long as you keep everything backed up regularly, you don't lose all of those family pics if your computer kicks the bucket. (Learn more about cloud storage on page 261.)

You might not be wild about trusting your precious memories to the cloud. Or maybe you think you need a backup for your backup. Consider an external hard drive that stays at home but is accessible via phone, tablet, or any other computer, wherever you are.

Check out a network-attached storage (NAS) drive — it's like having your own personal cloud. An NAS is essentially a server that resides in your home, connected to the Internet. There's no monthly fee, and you can sync and share across all of your devices. Many work with both Apple and Windows products.

Protect your hard drive from a fatal crash

It's as inevitable as death and taxes — one day, your computer's hard drive will stop working, and it will take every photo, letter, song, and file down with it. That wouldn't be so catastrophic if you knew when your hard drive would bite the dust, but you don't. It could happen today, or it could happen five years from now.

The best insurance policy is to create copies of all your files and save them to an external hard drive, a little box that plugs into your computer.

In Windows 10. The good news is, you don't have to copy every file and folder by hand. Windows 10 will do it for you, automatically, every day so that you never even have to think about it.

Start by plugging an external hard drive into your computer with a USB cable. Type "file history" into the Search box on your Taskbar and click *File History (Control panel)*. This opens the window where you set up and schedule your backups. Your computer should find the external drive you have plugged in, and show it here.

You'll have to turn on File History the first time you create a backup. To do that, click *Turn On.* Windows 10 will immediately begin to back up your files. You don't have to do a thing. When it finishes, you will see the phrase *Files last copied on* followed by the current date and time. Now if your hard drive crashes or if you permanently delete an important document, you'll have a copy tucked away safely.

Windows 10 will automatically back up your files to the external hard drive once an hour, if your computer is turned on. You can change that frequency, if you like. Click *Advanced Settings* on the left side of the File History window. Here you can change the backup schedule and choose how long the computer keeps old backups.

For Macs. Once you have a neat feature called Time Machine set up, you'll always have a backup copy of your hard drive on an external drive. As soon as you plug it in, your Mac will probably ask you if you want to use it for Time Machine. Click *Use as Backup Disk.* If it doesn't take you to the Time Machine settings, go to > System Preferences > Time Machine, and click to turn it on.

Got a laptop? You can still use Time Machine. It will take "local snapshots" that are stored on the laptop when you're away from the external hard drive, then merge them with the existing backups when you plug in again.

On the cloud. Cloud-based services can also back up your files, folders, and photos. Instead of residing on an external hard drive, they live on servers owned by the company you choose. OneDrive, Dropbox, and iCloud are a few examples. Go to page 261 to learn more about cloud services.

Pare down your password use

Isn't it enough that we have to remember a passel of passwords for practically everything? You need one just to access your own computer. It makes sense if you're at work, travel with a laptop, or share a computer with people you don't want reading your email. But if it's your own personal computer, you might consider getting rid of its password protection.

On a PC. Head to > Settings > Accounts > Sign-in options. From the *Require sign-in* drop-down menu, click *When PC wakes up from sleep.* That will buy you some time if you're quickly stepping away. If you never want to have to sign in again, click *Never.* You have to decide if you're willing to compromise the convenience for a bit of safety.

On a Mac. Go to > System Preferences > Users & Groups. Click your name, then *Change Password.* Put in your old one, but leave the *New password* and *Verify* fields blank. No more password necessary.

Keep your webcam from becoming a spycam

If your computer has a webcam, beware — someone could be watching your every move right now, secretly recording what you're doing in your own home.

When reports of webcam hackers first surfaced, many dismissed it as a kooky conspiracy theory at best. Turns out it's easy to access a webcam, and it happens more often than you think. Plus, hackers can remotely turn off the green light that's supposed to indicate when your webcam is on.

Is it likely that you have a spy? Not at all. But it's easy to make sure if someone did try, they'd have nothing to see.

- If you have a plug-in webcam, just keeping it unplugged when you're not using it is enough.
- You could try disabling your built-in camera, but this could result in bigger problems if you're not sure what you're doing. So the easiest and least expensive solution is simply to cover up the webcam. Buy cling-on, clip-on, magnetic, or stick-on accessories, some of which have covers that slide open.
- For a really low-cost solution, put a sticky note or a piece of electrical tape over the camera.
- Keeping your operating system and anti-virus software up to date will also reduce the chances of hackers using malware to access your webcam.

Protect your computer with anti-virus software

It's a tragic tale of woe: a computer so infected by malware — software such as a virus or spyware designed to attack computers — that a professional has to fix it. Fortunately, most malware can be stopped using a combination of anti-virus software and common sense.

Windows users. Windows computers come with Defender built in, a free, anti-virus software that automatically updates your machine, but the catch is you have to turn it on first. Go to ⊞ > Settings > Update & Security > Windows Defender, and turn on *Real-time protection*. You can also turn on *Cloud-based Protection*, which will send Microsoft a note when Defender finds malware, and *Automatic sample submission*, which will send Microsoft malware samples. These settings are meant to keep Microsoft in the know about the latest attacks.

Defender scans your hard drive periodically, but you can also perform a scan whenever you want. Go back into Windows Defender, scroll to the bottom, and click *Open Windows Defender*. Pick a scan option (*Quick*, *Full*, or *Custom*), and click *Scan now*.

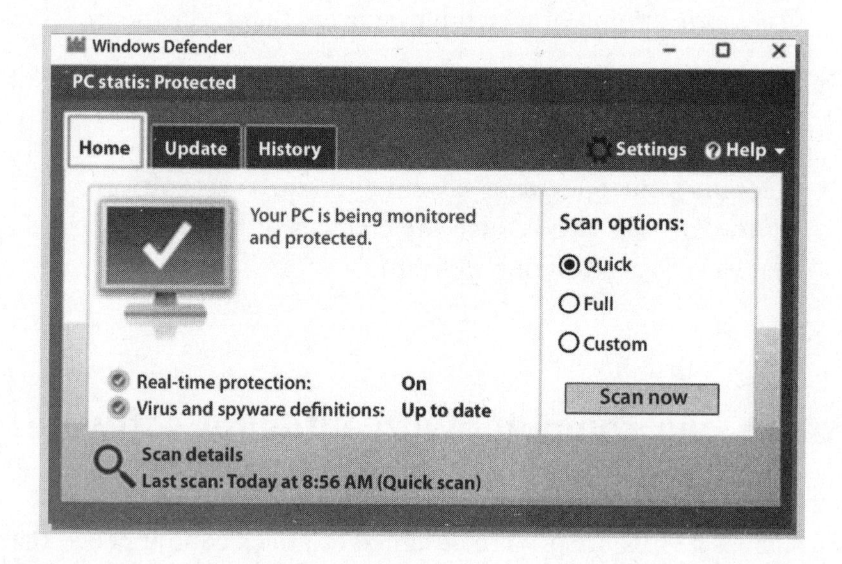

If Defender finds something suspect, an alert will pop up and the program will quarantine the virus before it can attack.

But how do you know if Windows Defender is enough? Only you can decide. Third-party programs can catch more viruses, but they may also slow down your machine and give you the occasional false alarm. If you're careful about where you click and what you download, you're probably fine with the built-in stuff.

For a greater level of security, Avast is one of the most popular anti-virus programs around. It's free, loaded with features, and has been a leader in the anti-virus game for 25 years. Visit *avast.com* for more info.

Mac users. You may have heard Macs don't get malware, because they have great built-in security systems. That may be true, but Macs still can get infected.

Apple doesn't include anti-virus software on its computers, but there's something simple you can do — keep your Mac's operating system up-to-date. When you get an alert to download a patch or update, do it as soon as possible, because updates often contain additional anti-virus protection.

It also pays to be suspicious of email links or attachments. When in doubt, don't click or open. Go to the source like the friend or the website who sent you the email, and investigate.

And if you want to take an additional precaution, Avast is also available for Macs.

Take a picture of your screen, it'll last longer

A picture may be worth a thousand words, but a screenshot may be worth a thousand dollars. Well, maybe not a thousand, but how about a hundred or more? That's because screenshots can help you troubleshoot computer problems. So how do you take a picture of your screen? It's easy — here's how to do it.

Use the Print Screen button or Snipping Tool app on a PC. A Windows computer offers a couple of options, either through a keyboard or a special app called Snipping Tool.

- On your keyboard, look for the PrtScr or PrintSc key. Press that button and you'll get a picture of your entire screen, saved to your clipboard. Then you can paste it anywhere you like.
- If you'd like to save the shot in your Photos folder, press ⊞ +Print Screen. It will be named Screenshot until you go into the folder and change its name.
- To get a screenshot of just the main window you have open, press Alt+PrtScr. Your Clipboard will save the image until you paste it into a file like a Word document.
- To "cut" an image as with a pair of virtual scissors, look for the Snipping Tool in the Start Menu. Snipping Tool lets you change the shape of the "pictured" area so you can edit it before saving. Choose from a *Free-form, Rectangular, Window*, or *Full-screen* snip.
- Snipping Tool also lets you capture up to five seconds of action on the screen, perfect for nabbing a pop-up menu that disappears quickly.

Use a shortcut or the Grab app on a Mac. You'll need three fingers to snap a shot on your Mac. Or just open the Grab app.

- To get the entire screen, press Shift+⌘+3. The screenshot will save on your desktop with a time and date stamp as its name.

- If you'd rather just get a small rectangular section, press Shift+⌘+4 and drag your mouse to capture what you need in the box. Release the mouse to save.
- Find Grab under Applications > Utilities. You'll see the Grab app menu across the top of your screen. Go to Capture > Selection. A pop-up box will let you drag around the mouse until you've gotten the area. Release the mouse and the captured image will pop open on your screen. Name it and save it.
- To capture a full screen, go to Capture > Screen. Click outside the resulting pop-up. If you need to delay the screenshot, go to Capture > Timed Screen, and you'll get 10 seconds to set up your screen first.

Access your computer from anywhere, anytime

You're on the phone with a friend who considers you her own personal tech support. You're happy to help, but it's frustrating when you can't even figure out what she's trying to do. Or maybe you're the one who needs computer help. Here's how someone can help you fix your computer without even touching it.

It's called "remote access," and it's a way to share screens with a friend or technician. Windows and Apple have their own remote access programs, but you can also download a third-party program. Which one you choose depends on your budget and how much access you want to give (or get).

Apple Remote Desktop is available for purchase in the App Store. You can select specific users with whom to share screens, and choose which tasks they can perform when they're accessing your computer. Everything is password protected, and both the sender and the receiver have to configure settings so the computers can communicate, but Remote Desktop walks you through the steps.

Microsoft Remote Desktop is available for both Macs and PCs, and it's free for personal use. If you have a Windows computer, you may already have it. Just click ■ and type "remote desktop" in the search box. Click *Remote Desktop Connection*, fill in the fields, and click *Connect.*

Phone and tablet versions are available, so you can access a computer even if you're not in front of yours. It's also useful if one of you has a Mac and the other has a PC.

Teamviewer is one of several free remote access programs. It's available for all operating systems and has gotten rave reviews. Setup is a snap. Both of you just download the software from *teamviewer.com*, and share the user IDs and passwords (uniquely generated each session). In seconds you can see your friend's computer and even remotely control it while she watches. Exit the program and the connection is gone.

Save electricity and battery life with sleep mode

When you're finished using your computer, what do you do? Nothing? Shut it down? Put it to sleep? Does it matter? Absolutely.

If you're going to be away from your computer for a few days, it's best to shut it down. With PCs especially, shutting down occasionally will help your computer run more smoothly. And it's a power saver. But that also means a bit of a wait when you turn it on and everything starts up again.

If you'll be coming back to your computer fairly soon — even the next day — you can just put your computer to sleep. It uses a little bit of power, and one tap on the keyboard wakes your computer back up and your programs are instantly active.

In Windows 10, go to ■ > Power, then choose from *Sleep, Update and shut down*, or *Update and restart*. Or go to ■ > Settings > System > Power & sleep, and pick how long you want your computer to stay awake before it goes to sleep when idle.

On a Mac, click , and choose from *Sleep, Restart...*, or *Shut Down....* Only click *Force Quit...* if your Mac becomes unresponsive. (To learn about setting up a sleep/wake schedule for your Mac, go to page 46.)

When you shut down your PC or Mac, make sure all your apps and programs are closed.

Apple OS X

Unleash the magic in your Mac

Find natural scrolling unnatural? Change it with a click

If you're new to Mac or recently upgraded from a later model, you may have received a shock. Sliding your finger down the mouse or trackpad makes the page go up. What? Shouldn't down mean down? Well, Apple has its reasons, namely, to make the Mac reflect its other touch-screen products — the iPad and iPhone.

Before you panic, know you can change these settings in a jiffy by clicking > System Preferences.

- For a Magic Mouse, click Mouse > Point & Click, and uncheck *Scroll direction: natural.* And while you're in there, explore other mouse preferences and make your Magic Mouse do whatever you want.

- If you use a trackpad, click Trackpad > Scroll & Zoom, and uncheck *Scroll direction: natural.*

And there you have it. You're old school again.

Discover the fun functions of your Fn key

Think of your Fn key as your fun key. That's because when you pair Fn with other keys, you get a whole new set of functions out of your keyboard.

For instance, where F11 turned down your volume before, Fn+F11 sends all your windows scurrying to the edge of the screen, so you can view your desktop.

Learn what else Fn can do with other keys and how you can use them by checking out the following table.

Fn keystroke	Action
Fn+Delete	Forward delete
Fn+11	Send windows to edge of screen
Fn+↑	Scroll up one page
Fn+↓	Scroll down one page
Fn+←	Scroll to the beginning of a document
Fn+→	Scroll to the end of a document

Hot shortcuts for your most in-demand commands

Your Mac has a lot of stuff going on. And finding your way around can be a hot mess. That's why Apple has introduced hot corners — another brilliant way to set up shortcuts on your computer.

Hot corners are the four corners of your screen, set up to perform some kind of command when you move your cursor to that corner. To set up your hot corners go to > System Preferences > Desktop & Screen Saver > Screen Saver. On the bottom right of the window, click *Hot Corners*. From the *Active Screen Corners* dialog, choose a command from the pop-up menu by each corner.

For example, set the top-left corner to *Put Display to Sleep*. Next time you need to leave your computer, you can hide your screen in an instant with a quick flick of your mouse in that top-left corner.

Do you always have a bunch of windows open? Set the bottom-right corner to *Desktop*. Whenever you need to clear your desktop, just move your pointer to the bottom-right corner and all of the windows will move out of the way.

If you keep triggering hot corners accidentally, you can add a modifier key or keys like ⌘, Shift, Option, or Control. For instance, with the *Hot Corners* pop-up menu open, press Shift and Command simultaneously, and select *Desktop*. (See example on next page.)

From now on when you move your cursor to the bottom-right corner of your screen, you'll have to press Shift and Command for your windows to scramble.

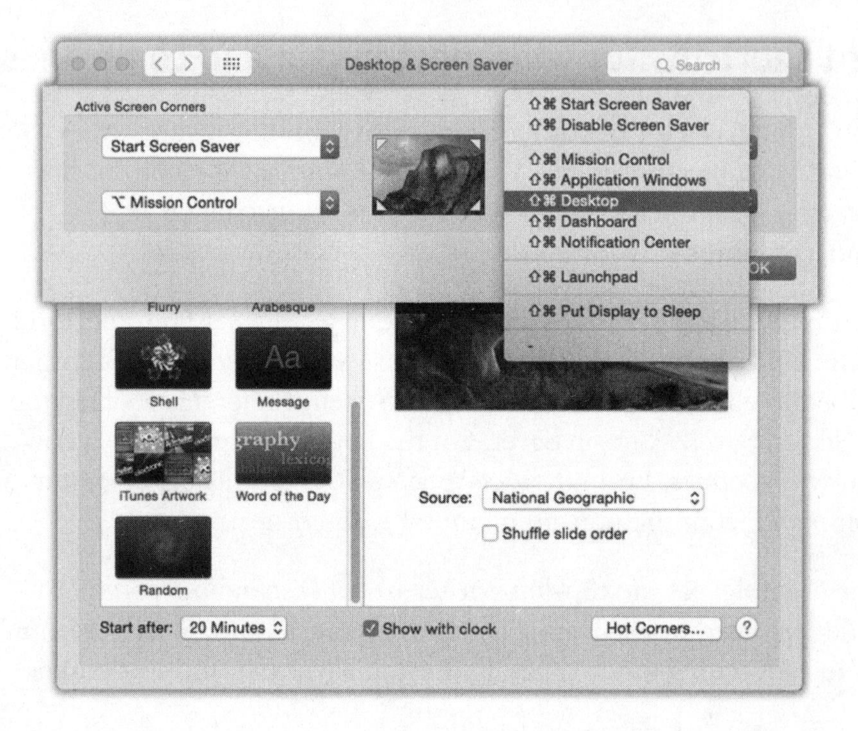

Teach your Mac to wake up and sleep on your schedule

You don't have to wait for your computer to wake up every morning or waste money and electricity by leaving it on all night. If you use your computer at the same time every day, you can make your computer start up right before you need it, and sleep or shut down when you don't. Set it up once and you never have to think about it again.

Scoot over to > System Preferences > Energy Saver > Schedule, and adjust the settings to your heart's content. Next time, your computer will be standing at alert, exactly when you're ready for it.

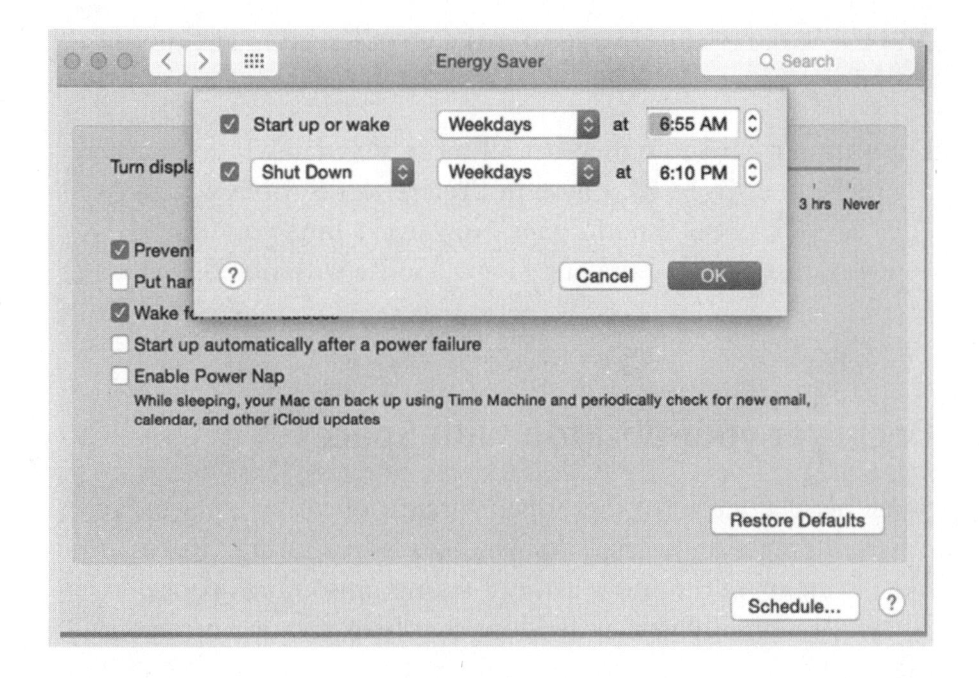

Magnify your screen and put your squinting days behind you

Is it just you or do screens seem to be getting smaller? Does that recipe say "pound the steak" or "pound of cake"? You can stop guessing. Your Mac can magnify your screen so you can make out every single letter, and keep your dinner guests happy.

Go to > System Preferences > Accessibility > Zoom. Pick between *Use keyboard shortcuts to zoom* or *Use a scroll gesture with modifier keys.*

What's more, you can turn your pointer into a magnifying glass by checking one of those two options, then changing the *Zoom Style* from *Fullscreen* to *Picture-in-picture.*

And check out *More Options* to make your screen image move with the pointer.

But wait, it's hard to remember all those shortcuts. If you have a Magic Mouse, go to > System Preferences > Mouse. Under *Point & Click*, check *Smart Zoom*. Now every time you want to zoom in, just double-tap your Magic Mouse with one finger.

Declutter your desktop with Spaces

Some people like all of their open applications in one place. For others, a tangle of overlapping windows is frustrating. Spaces are like having extra monitors without taking up so much room. They're essentially virtual desktops that help you organize your Mac life.

Mission Control is where Spaces live. And it's even more user-friendly with OS X El Capitan. Press the Mission Control key (usually F3) and you become the all-seeing user. Move your pointer to the top of the screen to take a peek at mini versions of your desktops. You get a visual on all the apps and windows open on your computer.

Create up to 16 spaces. It's easy to create new spaces. In Mission Control, click the plus sign in the top-right corner. Or drag a window to the plus sign to create a new space containing that app. You can also make an app full screen, which makes it a temporary space, but it doesn't count toward your 16 available spaces.

If having more than one desktop gets confusing, change the background of each of them so you always know which one you're viewing.

Nix the extras. If you go a little space happy you can always delete them. Go to Mission Control and move your pointer to the top of the screen to view your desktops' thumbnails. Press Option and click the X in the top-left corner of the space. Or hover over the space to reveal the X.

Switch spaces like a pro. You have several ways to switch between spaces.

- Swipe left or right with two fingers on a Magic Mouse (set up in mouse preferences) or with three fingers on a trackpad. Swipe slowly to just peek into the next space instead of switching all the way.
- Go to Mission Control and click the desktop image of the space you want.
- Use Control and your arrows to move left or right to other spaces.
- Press Control and the Desktop number. You'll have to set this up in > System Preferences > Keyboard. Click Shortcuts > Mission Control > Switch to Desktop.

Arrange Desktops in the order you want. You can always rearrange your spaces in Mission Control. Just drag the desktop images to the order you want them to appear. If you want Spaces to stay in this order, go to > System Preferences > Mission Control and uncheck *Automatically rearrange Spaces based on most recent use.*

Pin apps exactly where you want them. Say you want Safari to always open on Desktop 4. Navigate to *Desktop 4* and right-click (or two-finger click) the app icon in the dock. Select *Options* and *This Desktop.* (See example on next page.)

If you use a program all the time, Finder for instance, right-click the icon and select *Options* and *All Desktops*. The program will show up in every space, so you don't have to switch spaces every time you need it.

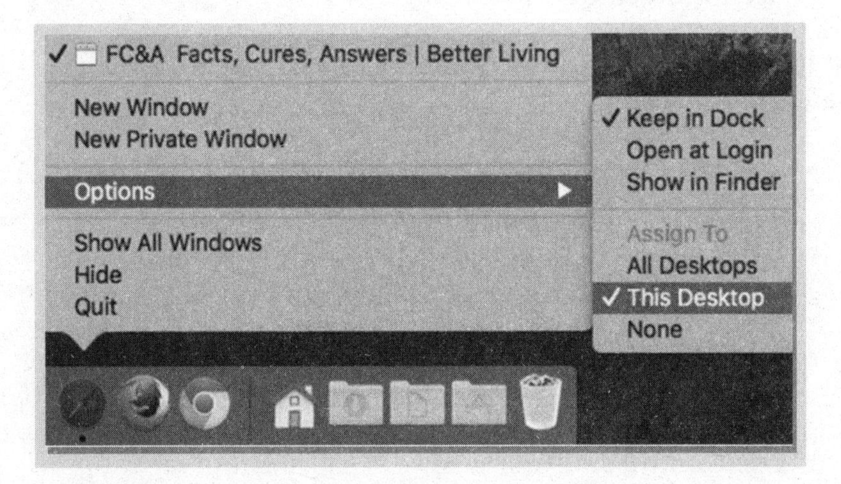

Move apps from one place to another. Maybe you want your calculator to open next to your spreadsheet, but it's on Desktop 4 and your spreadsheet is on Desktop 5. Here are two simple ways to scoot that calculator over.

- Move it to another space by dragging its title bar to the right or left edge of the screen. Hold it there for a moment and the window will slide into the next space.

- Or just grab the window's title bar and drag it to the top of the screen. You switch instantly to Mission Control, where you can drop it into any space you'd like.

Take apps full screen to get rid of distractions

It's the 21st century, and you're bombarded by information overload. As if you need any more distractions, you have menus, scroll bars, and status bars cluttering your screen. If you need to work on something with minimal disturbance, switch to Full Screen and only eyeball what you're working on.

Full Screen mode is available in most Apple programs. Going Full Screen will open the app in a new Space. You have several ways to get there.

- You can click an app to view it Full Screen if the green button at the top left of the program window shows a double arrow when you hover over it. Once you're in Full Screen, the menu bar will appear when you move your pointer to the top of the screen. You can click the green button again to exit Full Screen.
- In the menu bar, go to View > Enter Full Screen. To exit, go to View > Exit Full Screen.
- Press Control+⌘+f. Press the keys again to exit or just press Esc.

To learn how to turn your Full Screen app into split view, visit page 24.

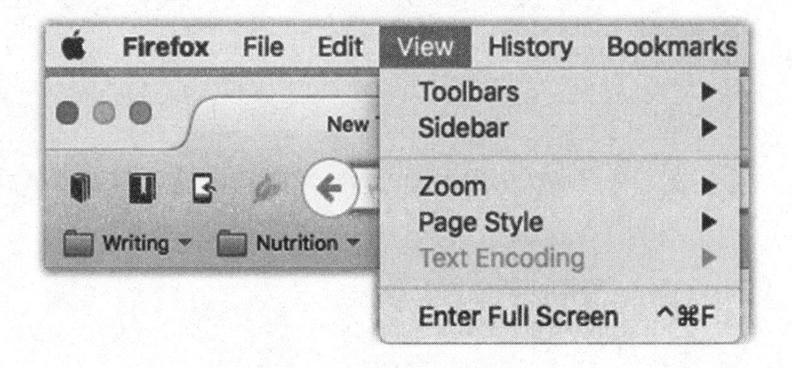

Finder tab-dos and tab-boos

Look near the top right of your Finder window. See that little plus symbol? That's right — Finder has tabs just like your Web browser. And if you don't see it, go to View > Show Tab Bar. That little plus sign adds all sorts of possibilities for viewing files.

- Open a new tab by clicking the plus sign. Or just press ⌘+t.
- Right-click (or two-finger click) the file name in the Path Bar located at the bottom of the Finder window. If you don't see it, click View > Show Path Bar. Select *Open in New Tab* from the drop-down menu.
- If you've already located a file you want to open in a new tab, you can also drag the file to the plus symbol. (See example.)

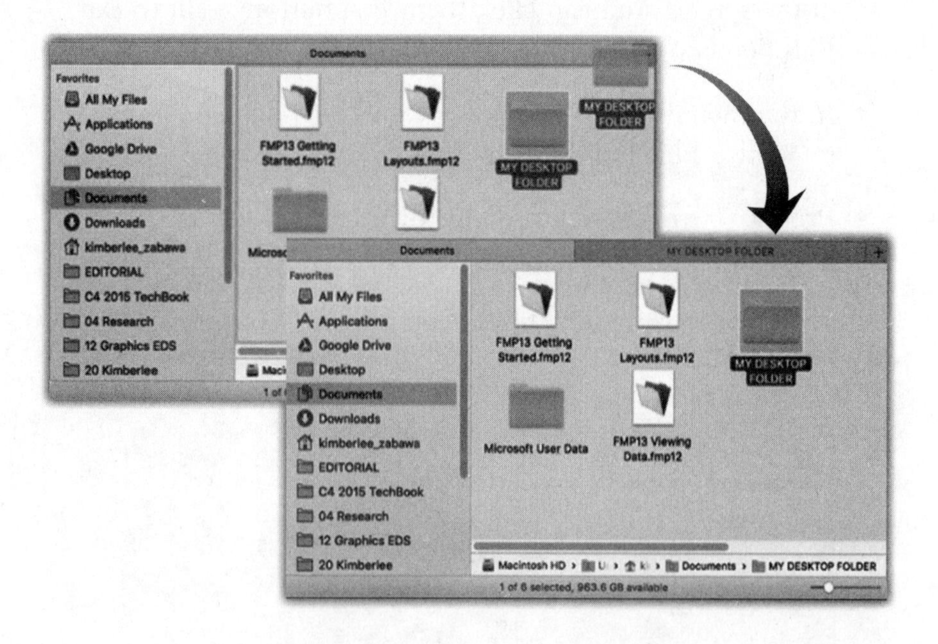

Watch what you're doing though. If you drag the file to an open tab instead of the plus symbol, you will switch to that tab and copy the file to the contents of that tab. Also, if you drag a tab out of your current Finder window, it will open in a new one.

Never get lost in Finder again

Winding your way through files in Finder can be so frustrating. Where are you? How did you get here? And most of all, how do you get where you need to be? Your Mac has two built-in solutions to help you find your way.

The title bar has a secret map of your past. Searching for a file can be like getting lost in a box of Russian nesting dolls. You could click your back arrow button. But here's a secret — you can right-click (or two-finger click) the name of the window (title bar) to view a hierarchy drop-down menu. Then click on the file you wish to return to, and snap! There you are.

As a bonus, Mac allows you to do this in other programs, too. Right-click (or two-finger click) the title bar to see exactly where that file is saved. You can click on a file name and jump to it in Finder.

The Path Bar is like a breadcrumb trail. You're in luck if you're the kind of person who hates asking for directions. You can reveal a Path Bar in the bottom of your Finder window that spells out the path it took to get you where you are.

Double-click a file name to jump back to that place. If you don't see the path bar in your Finder window, click View > Show Path Bar. (See example on next page.)

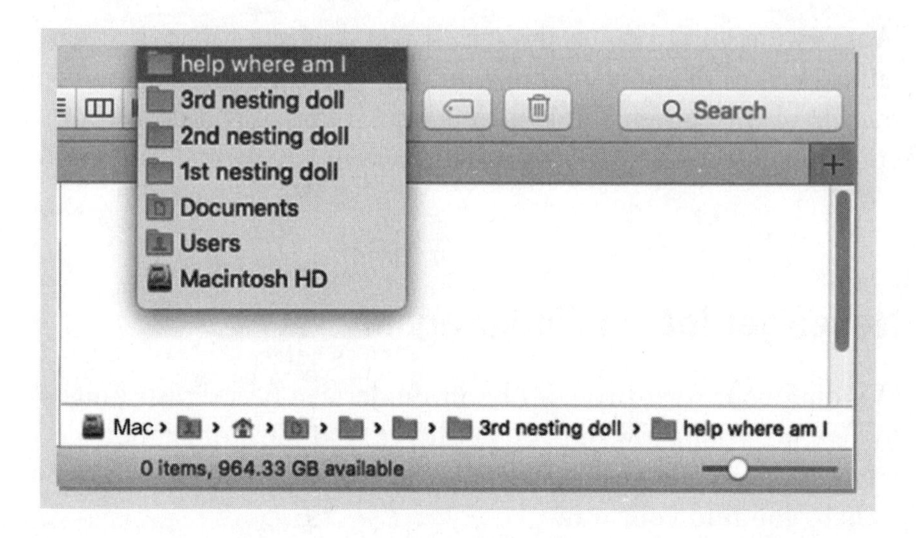

Peek into a file without opening it

Finding files can be a hassle — especially if you have to open each one to see if it's the right one. But with Quick Look, you can get a sneak peek of a document, image, or sound without waiting for the program to wake up. And impress your friends with your lightning-fast speed.

Quick Look is already built into your Mac. Simply select the file you want to check out, then press the space bar. In a flash, this quick and handy tool let's you see a preview of the file with several useful options.

- Select multiple files, and flip through them using the arrows at the top of the box or by swiping your Magic Mouse with one finger.

- Scroll within the document itself.
- Share the file through Mail, Messages, social media, and more. Look for the little box with the upward-pointing arrow on the top-right corner of the window.
- Open the file straight from Quick Look.
- If it's a movie or sound, you can even watch or listen to it.
- Toggle to full screen to watch a slideshow of your favorite pics.
- And you don't have to close your current preview to look at another file. Just click on a different file and the preview will change instantly.

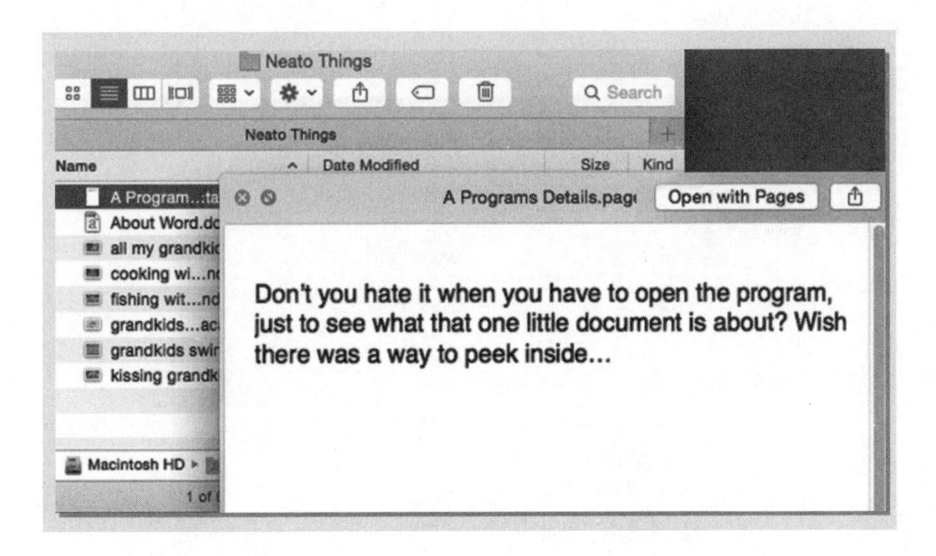

Create a PDF in 3 easy moves

PDF. The Portable Document Format. Named so because it can be shared and opened on almost any device. And the convenience doesn't stop there. Creating a PDF on a Mac is as simple as 1, 2, 3.

Create from Print. You can make a PDF from almost anywhere — PDF your flier in Pages or your new favorite recipe on the Web. Whatever program you find yourself in, you can usually follow these easy steps.

1. Press ⌘+p.
2. Instead of clicking *Print*, choose *Save as PDF* under the drop-down menu in the lower-left corner.
3. Click *Save*.

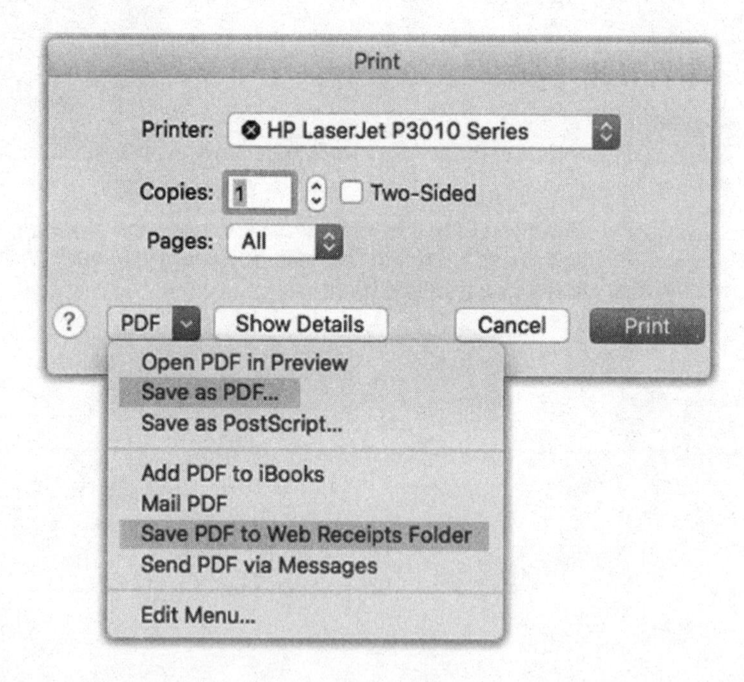

Your PDF is ready to be shared, printed, or even marked up in Preview.

Save as a PDF. Some programs allow you to save documents as a PDF straight from the *Save* dialog box. For example, if you are in Word, you can select *PDF* from the *Format* drop-down before clicking *Save*.

Never lose another receipt

You just bought a new blender online, but before you print another receipt you're probably going to lose anyway, check out this way to cut down on paperwork. Zip over to the *PDF* drop-down menu on the bottom-left of the print dialog box and select *Save PDF to Web Receipts Folder*. (See example on previous page.)

Your receipts will be organized in Home > Documents > Web Receipts. Then you can flip through them and print at your leisure. You may even learn you just bought that same blender last week.

Insert an accent mark without leaving your keyboard

So you need an accented letter again. Maybe you're emailing your friends to meet you at the café this evening, and you want to sound extra European — "cafe" just won't do. Great, but where are those special characters? Look no further than your keyboard.

Just like on an iPhone, all you need to do is hold down the letter you want accented and — voilà — you get a pop-up panel of accent options. Click on the accented letter or just type the number below the letter. Fantástico!

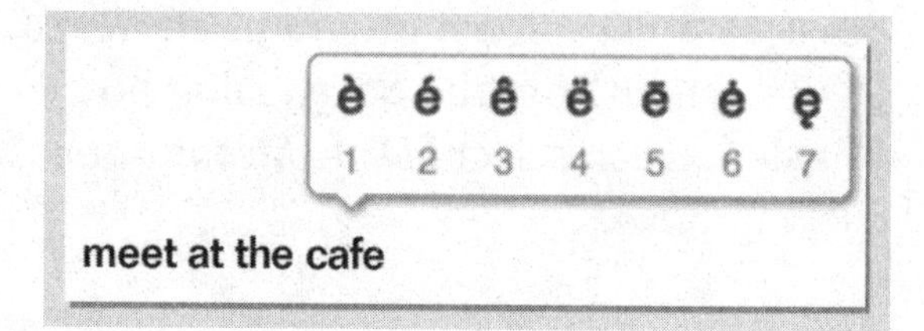

Simple shortcuts to all your emojis and symbols

You may not use special characters every day. In fact, you probably only dig through your menus often enough to figure out where they are, with just enough time in between to forget where you found them.

One way to insert special characters is to go to the Menu bar and select Edit > Emoji & Symbols. But you can also put the Viewer icon in your Menu bar so your special characters are just a click away. To do this, go to > System Preferences > Keyboard. Under the *Keyboard* tab, check *Show Keyboard, Emoji, and Symbol Viewers in menu bar.*

All symbols in one place with Character View. Click the Viewer icon in your Menu bar then select *Show Emoji & Symbols.* This will open a window full of every symbol you could ever want. With categories such as *Emoji, Arrows, Math Symbols*, and *Latin*, you'll be like a kid in a candy store.

Keyboard View makes it easy to remember shortcuts. It's hard to remember the keyboard shortcuts for all those symbols. To see which symbol each key can produce, click the Viewer icon in your menu bar then select *Show Keyboard Viewer.* You will get a mini on-screen keyboard. Press the modifier keys ⌘, Option, Control, or Shift to see which symbols they produce.

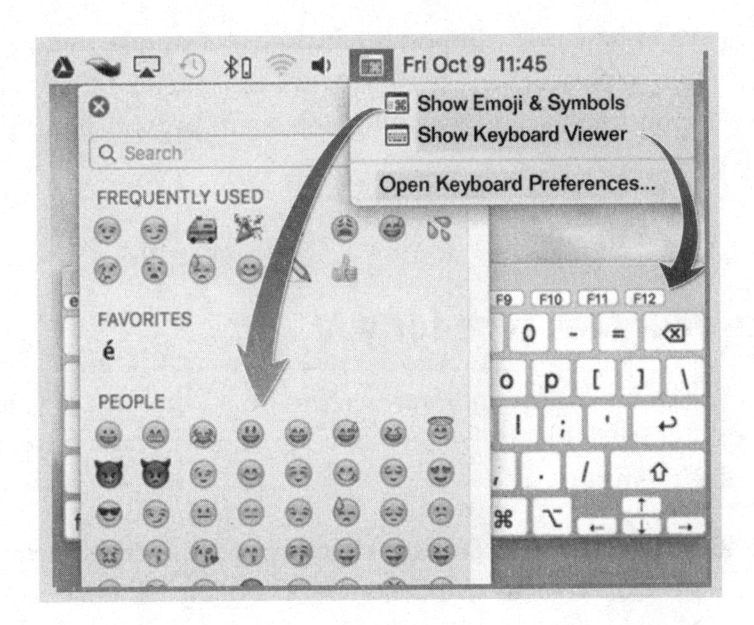

Auto-complete — it's not just for your phone

OK, so you weren't the fourth-grade spelling bee champion. And you're still not quite sure how to spell "convenient." That's OK — your Mac won't judge. It has a built-in feature just for you.

You're probably familiar with auto-complete on your phone. Your Mac has a similar feature. Begin typing the word then test these keys to see which one auto-completes on your keyboard.

- Esc or Option+Esc
- F5
- ⌘+period

A list of possible words will pop up. To select a suggestion, press Return, Tab, or Space. Or use the arrows to navigate up and down the list. If you don't see the spelling you want, press Esc or just keep typing.

Make your Mac type for you

Now you can tell your computer what to do — literally. If your fingers don't work like they used to, your Mac will type for you with voice typing.

Go to > System Preferences > Dictation & Speech > Dictation to turn on *Dictation* or to *Use Enhanced Dictation* offline.

When you're ready to put your Mac to work, click where you want the text to appear and press the Fn key twice. You will see a microphone and hear a beep. That's your cue to start talking. Speak normally — no need to talk slowly or dramatically. The computer understands much more than your dog.

Also, don't forget to speak your punctuation. For example, "Dear editors (comma) (new paragraph), Thank you so much for writing this book (period). You have no idea how much it has helped me (dash) my kids think I'm brilliant (period)."

The secret to making your Mac read to you

Want your computer to read your email while you make breakfast? What about tell you the latest news as you brush your teeth?

Many Apple programs — including Safari, Mail, TextEdit, Messages, Stickies, and Pages — have a feature that allows your Mac to read out loud. Highlight the text, right-click (or two-finger click) inside the program window, and choose *Speech* from the pop-up menu. Select *Start Speaking.*

If you have a thing for accents, you can customize your *Text to Speech* feature. Say you want your Mac to speak to you as an Australian bloke named Lee or a Spanish señorita named Monica, go to > System Preferences > Dictation & Speech > Text to Speech. Pick a voice from the drop-down menu under *System Voice.* After your Mac loads your voice preference, check *Speak selected text when the key is pressed.* Then click *Change key* and pick a key combination.

Highlight what you want read and press your shortcut key combo. Your Mac talks back in the voice you picked. Though if you're a parent, you may already have enough of this at home. To turn off *Text to Speech*, simply press the key again.

Send a file instantly — without ever opening email

Instant messaging has been revolutionary. No more waiting for your message to send. No more refreshing the screen until an email shows up in your inbox. When you chat through Apple's Messages, you have a real-time conversation. But what some people don't know is you can do more than that.

If you're already messaging your best friend, there's no need to pull up Mail to send those photos you're chatting about. You can send pictures, audio clips, videos, documents, and even entire folders through Messages.

Just drag the file you wish to share and drop it in the chat window. Or press Option+⌘+f to open the dialog box and select the file you want to send.

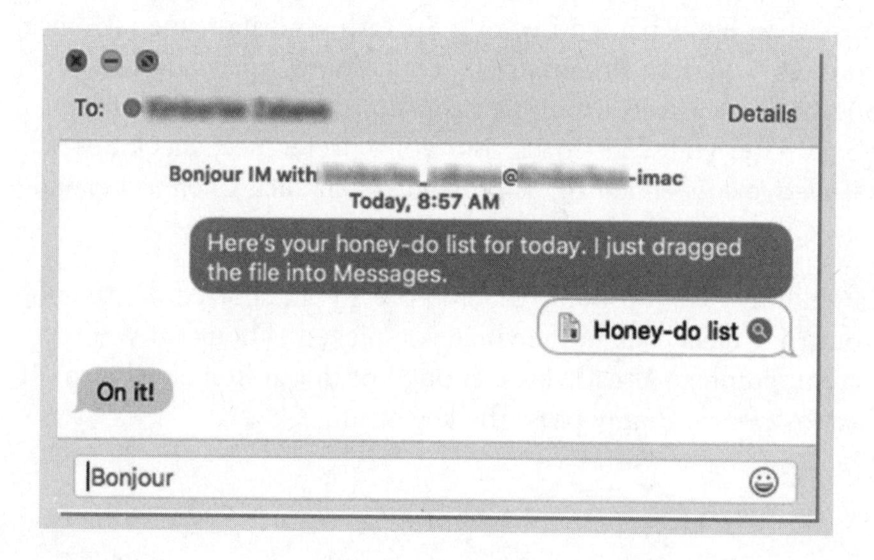

All of the files you send and receive through Messages will be available, even when the conversation is over. To see these files, go to Window > File Transfers.

Long distance and video conferencing – totally free and a lot more fun

Long distance calls and video conferencing can get expensive. And finding the right plan and program is a hassle. But just when you thought you'd be cut off from the rest of the world forever, you learn Messages gives you two options that will make your communication needs a lot easier and a lot more fun.

Free long distance with audio chat. As long as you're connected to the Internet, you can chat anywhere. Audio chat supports conversations of up to 10 friends with the party-line feature. So when you finally take that dream vacation on the other side of the world, you can still get in touch with your friends and brag all you want. Just go to Messages and click the little phone icon at the bottom of your Buddies window.

Simple video conferencing with video chat. If you want a face to go with that audio, you and up to three of your friends or colleagues can talk face to face with the video chat feature. Just open Messages and click the video chat icon, found at the bottom of your Buddies list. It looks like a little video camera.

2 ways to send a text straight from Messages

When you're working on your Mac, you don't want to juggle several devices at once just to send a text. Apple makes keeping in touch easy by giving you two ways to text right from your Mac using your favorite instant messaging app — Messages.

Message your friends without using your data plan. iMessages allows you to move from your iPhone to iPad to iPod to Mac without missing a beat, as long as you have Internet. So if you start a conversation on your Mac, then decide to take a walk, you don't have to end the chat. Just scoop up your iPhone and pick up where you left off.

Plus, characters aren't limited to 160 like standard texts, and you can also share photos, movies, documents, and large files — with more than one person at once. So if you're trying to make a family decision, text the whole gang. The best part is you don't have to use your data plan because you're not sending messages through your cellphone network.

On your Mac, go to Messages > Preferences > Accounts, sign in with your Apple ID, and click *Enable this Account*. You will also need to turn on iMessages on your iPhone in Settings > Messages.

Send a regular text — to anyone. Like iMessages, you can send another type of text in your Messages application. But that's where the similarities end. This feature sends a standard text message, using your data plan.

Why would you want to do that? Well, you can only send iMessages to people who have iPhones. For everyone else, *Text Message Forwarding* is the way to go. To set this up, your iPhone and Mac have to be signed in to the same iCloud account as well as connected to the same Wi-Fi network.

- On your iPhone, go to Settings > Messages and tap *Text Message Forwarding*. The name of your Mac should show up and you can turn on the switch. A box will pop up, asking for a code you will find on your computer screen. (See example on next page.)

- On your Mac, open *Messages*. The code will appear and you should type the code into the box on your iPhone. You can also set up texting on your iPad or iPod.

To send a text, open Messages and type the name or number of your contact. Messages will let you know the person is a Text Message contact and not an iMessage contact.

Answer and make calls right from your Mac

It's easy to miss a phone call when you're in the zone, working on your computer. But now, you can set up your Mac so you can make and answer calls without picking up your phone — your calls show up right on the screen.

You need to do several things before you call your friends and neighbors to share the good news.

- Sign your Mac and iPhone into the same iCloud account.
- Connect both devices to the same Wi-Fi network.
- Turn on Bluetooth on both devices. They must be within Bluetooth range, which is about 30 feet.
- Set up your Mac by going into your *FaceTime Preferences* and checking *Calls From iPhone.*
- Turn on FaceTime to allow cellular calls on your iPhone in Settings > FaceTime.

Answering a call on your Mac is simple. You don't even have to have FaceTime open. When someone calls you, a notification will pop up with the name of the caller. Click either *Accept* or *Decline.*

You can even phone a friend from your Mac. In FaceTime, type the name or phone number of the person you want to call in the *Enter a name, email, or number* field. Then, click the phone icon next to the name or number. From the drop-down menu, choose the number you want to call, and your Mac will begin calling.

You can also select a number in your Contacts or an app such as Safari, and FaceTime will start dialing. Pretty cool, right?

Start a task on one device, hand it off to another

You're a busy person, and sometimes you start a project on one device, then have to dash off, leaving that project unfinished. But now you can pass unfinished projects from your Mac to your other Apple devices and vice versa — wirelessly and automatically — via Handoff.

Handoff only works in one app at a time, but you can use it with many different programs, including Maps, Reminders, Calendar, Contacts, Safari, Notes, Numbers, and Pages. To set it up, make sure all devices are signed into your iCloud account.

You'll need to turn on Bluetooth on the devices you want to use. Your devices must be sitting within Bluetooth range of each other, which is about 30 feet.

- On Mac. Go to > System Preferences and click *Bluetooth.*
- On iPhone, iPad, or iPod. Tap Settings > Bluetooth. Next, make sure Handoff is enabled on all devices.
- On Mac. Go to > System Preferences > General and check *Allow Handoff between this Mac and your iCloud devices.* If you don't see this option, your Mac doesn't support this feature.
- On iPhone, iPad, or iPod. Turn on *Handoff* in Settings > General > Handoff & Suggested Apps.

Take your work mobile. If you start reading a great article at *fca.com* on your Mac and want to read it on the go, just grab your iPhone and watch the magic happen. A Handoff app icon, in this case Safari, will automatically show up on the bottom of your device. Swipe up to pick up right where you left off.

You'll also see this icon when you double-press the Home button. Swipe left to see the Handoff app, then tap and go.

Move to your Mac. If you need to hand off your work to your Mac, the process is just as simple. When your devices are near each other, the Handoff icon will show up on the left side of the Dock. It also appears with your other apps when you press ⌘+Tab to switch between apps. Click the Handoff app icon to continue your work.

Share files effortlessly with no Wi-Fi

Picture this. You're at the airport and you just found out your flight's been delayed. You and your travel buddy grudgingly pull out your Macs to kill some time, and you begin flipping through old photos. That's when your friend sees one of the two of you at that big football game, and he has to have it.

Don't go through the trouble of connecting to finicky airport Wi-Fi and attaching photos to an email. AirDrop is a simpler way to share files between Apple devices with no tricky set up — and best of all, no Internet connection.

Open your Finder window and select Go > AirDrop from the Menu bar. Or just press Shift+⌘+r.

Any files you can transfer using the share icon (upward pointing arrow in a box) can also be sent via AirDrop — photos, videos, PDFs, and more. Just drag the file to the person's icon on your AirDrop window. They'll find it in their Downloads folder.

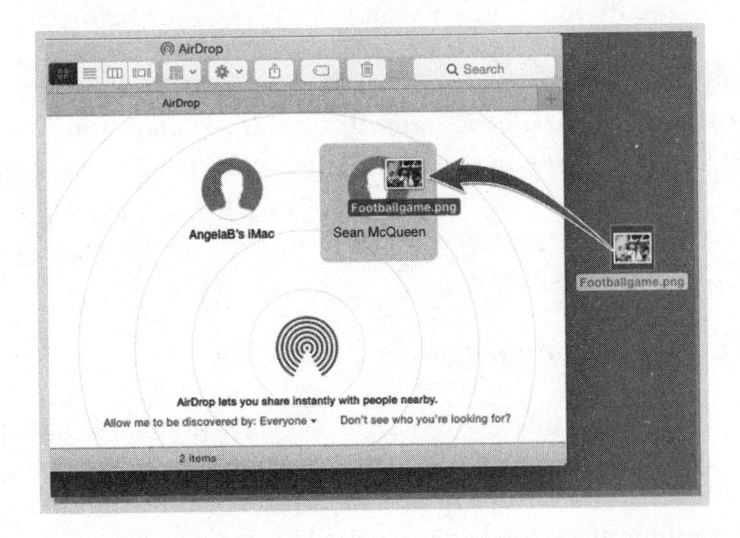

AirDrop is fairly new and still has a few glitches, especially with older devices. If you're having trouble getting AirDrop to work, remember you need to have Bluetooth and Wi-Fi turned on, but you don't need access to the Internet as the computers use a secure direct link.

So whether you're on the beach, at the pool, or relaxing in the hotel, as long as you are closer than 30 feet from your friend, you can share those pictures of you skipping work and living it up on your vacation.

Sharing media is a breeze with Family Sharing

Sharing — sometimes it can be a pain in the neck. But if you've been doing it since you were old enough to hold a toy, it just comes naturally — even when it comes to technology. And with Apple's Family Sharing, you can share like never before.

Family Sharing is free. Just go to > System Preferences > iCloud, sign in to *iCloud*, and select *Set Up Family*. You can have up to six family members under your account. Set it up with one credit card and only pay once for books, music, apps, and more. So if your daughter wants to watch a movie, all she has to do is hit play — on her own screen.

And that's not all Family Sharing is good for.

- You can set up *Ask to Buy* for kids and grandkids under the age of 13. If they try to buy something, they have to get permission first. You will receive an alert, and you can accept or decline the purchase request right from your own device.
- Explore an up-to-date photo album for the whole family. You may not live in the same state as everyone in your family, but

you can feel like you do with a shared photo album. It's a perfect way to still be involved with the family.

- Family Sharing syncs your Calendar and Reminders. That way everyone is always in the know. "I forgot today was family photo day" is no longer an excuse.
- Keep tabs on the people you love most. You can set up your device to share your location with the rest of the family automatically. Plus, if someone loses their phone or other Apple device, the whole family can pitch in.

You should remember everyone on your Family Sharing account will be able to see your purchases unless you choose to hide them — not that you'd have any reason to do that.

Move your old info to your new Mac — effortlessly

Yes — you finally got a new Mac! But now you have a brand-spanking new computer, complete with that new computer smell, and an old computer sitting pretty with all of your important files. You could manually move your files, but what about all those cool apps you paid for? Don't worry — all your accounts, settings, files, and other information won't be lost. You can transfer all to your new Mac with one of Mac's handy transfer assistants.

Migrate apps now with Setup Assistant. The first time you use your new Mac, the Setup Assistant wizard will hold your hand through the transfer process as you move all your old information to your new Mac. Just follow the steps to give your good ol' apps a brand new home.

Transfer later with Migration Assistant. If you don't migrate your apps and documents with the *Setup Assistant*, you can do it later. Go to Applications > Utilities > Migration Assistant. You can transfer information from a Mac, Time Machine backup, startup disk, or Windows PC.

When using Migration Assistant, keep in mind choosing to replace the existing user on your newer Mac will override that account's Home folder. Only choose this option if you don't have important information in the newer computer's Home folder.

Forgot your password? 2 ways to never get locked out again

You forgot your password — again. It's understandable. You have a million and one passwords to remember, and sometimes they just slip your mind. Unfortunately, this password holds the key to unlocking your Mac — and everything it holds. Luckily, you have two ways to get back in that don't involve throwing your computer across the room.

Trick No. 1 — the password hint. Apple has a wonderful safety net built-in for situations like this. If you fail to provide the correct password three times in a row, you will be offered a hint. The downside is if you haven't already set it up, it won't work.

To set it up, go to > System Preferences > Users & Groups > Password > Change Password. In the pop-up box, be sure to fill in the *Password Hint*.

Trick No. 2 — the magic reset. If Trick No. 1 doesn't work, you still have another chance. Again, you have to set this up in advance.

Go to > Users & Groups, and check *Allow user to reset password using Apple ID.* Next time you start to panic after your third failed attempt, you'll see a message that says, *If you forgot your password, you can reset it using your Apple ID.* Take a deep breath, and click the arrow-in-circle icon to reset the password with your Apple ID.

Rewind to an earlier version of your document in a snap

If you've ever created a document, you know how easy it is to lose important information.

Many programs automatically save your docs while you work. But did you know versions are also saved when you open, duplicate, lock, or rename a document? Your Mac keeps these snapshots around in case you need to go back to a previous version.

If you are working in Pages, for instance, and you need to see an earlier version of your document, go to File > Revert to > Browse All Versions.

In this view, you will see your current document on the left side of your screen and previous versions on the right. You can scroll through and compare versions by clicking the arrows, title bar, or timeline notches on the right side of the screen.

When you find the version you want to work with, click on it.

- To revert to the previous version, click *Restore.*
- To create a copy of the version in a new document, press Option and click *Restore a Copy.*
- To delete the version, reveal the menu by moving your pointer to the top of the screen. Select File > Revert To > Delete This Version.
- If you get cold feet, click *Done* to leave the document as is.
- You can also copy the text within a snapshot document and paste it into your current document, say if you wanted that really awesome paragraph you accidentally deleted a couple saves ago.

Sometimes you may see different options under File > Revert To. This happens if you explicitly save a version by clicking File > Save or Save As. In this case, you can select either File > Revert To > Browse All Versions or File > Revert To > Last Opened, Last Saved, or Previous Save.

Repair mysterious glitches with your Mac's First Aid kit

Even though your Mac may seem like a magic genie, it's still a computer, which means sometimes your programs goof up and leave you mystified. Luckily, your Mac comes with a program that can demystify many computer glitches and make your Mac run smoothly again.

Disk Utility is located in your Applications folder. It's a first aid kit for your computer — and the only thing you have to do is click a button. Well, five buttons. First open your Apps folder and go to Utilities > Disk Utility. Select *First Aid* and click the name of your hard drive, which is usually Macintosh HD. If it already names your hard drive, simply click *Run*.

Your computer will work its magic and with any luck, fix whatever is causing your problem.

You're only 2 keys away from a calculator

Everyone knows you're a math genius. In fact, you only dig through your apps folder for the calculator so it doesn't feel left out. But if you ever want to double check your calculations in the blink of an eye, just use Spotlight.

Click the magnifying glass at the top-right corner of your screen or press ⌘+space bar to bring up the Spotlight search box. Type or paste your math problem. You don't even have to push return to find out, or ummm confirm, that "sqrt(676) = 26" and "408/24*17 = 289."

The new Notes has noteworthy upgrades

The El Capitan operating system has given Macs a subtle facelift. But nothing has been transformed quite like the Notes app. Notes used to be a convenient place to jot down ideas, but it wasn't useful for much else — especially if you wanted a visual masterpiece with nice fonts. But the updated Notes comes with a lot of cool new features.

- **Add attachments**. In many apps like Safari, you can add attachments to new or existing notes with the app's share button (upward pointing arrow in a box). You can also add an attachment from within Notes by going to Edit > Attach file.

- **Open emails**. Now you can drag an email into your note to create a link that will open your message. You'll see the subject line and date before you click on it.

- **Change fonts**. This version of Notes has more options for you to customize your note. Change the font in Format > Font > Show fonts.

- **Make a to-do list**. Notes allows you to create checklists two ways — by clicking the circle with the check in the middle or going to Format > Checklist. You can actually check off items as you go.

- **Take a photo or video**. You can take a photo or even a video from Photo Booth. Then from Notes, pull up the photo browser (two overlapping rectangles icon) and just drag the photo or video straight to your note.

Changes are updated on all your devices through iCloud so you can take your notes with you wherever you go.

Microsoft Windows

Mastering your PC

The one-click way to delete unwanted programs

Bloating should only happen after a good meal with good friends — not on a brand-new computer. But every day, unsuspecting customers like you buy computers loaded with programs you don't need, didn't ask for, and can't on your life figure out how to remove. The struggle is over.

Windows 10 lets you uninstall programs directly from the Start Menu (⊞). The program may be sitting in plain sight on the Start Menu, but if not click *All Apps* on the bottom left and scroll until you find it. Right-click the program's name and select *Uninstall* from the pop-up menu. Then wave goodbye as it disappears.

Organize the Start Menu to meet your needs

It's like nothing you've ever seen, but don't be afraid of the Windows 10 Start Menu. Use the tips to make it work just the way you want.

Take away tiles. Are you a curmudgeon? Do you long for the old days of Windows XP, before all these new-fangled "tiles" showed up on your Start Menu? Then it's time to celebrate, because you can make any tile go away for good. Simply click ■, right-click the offending tile, and choose *Unpin from Start Menu.*

Add more, not less. Or maybe you like change because it keeps your mind sharp. In that case, add more tiles. Click ■ and select *All Apps.* Scroll until you find an app you want to see front-and-center each time you open the Start Menu. Then right-click its icon and choose *Pin to Start.*

Stay organized. One wild night of tiling can create complete chaos. Impose order by dragging and dropping your tiles into any arrangement you want. You can even create groups with labels like "Things I have to do but don't want to do" and "Things I'd rather be doing." Grab a tile and drag it away from the existing groups until a horizontal bar appears. Drop the tile to create the new group, then add other tiles to it. Now assign the group a name. Click the blank area above your new group, type in the box that appears, and press Enter.

Shrink tiles for small screens. Not everyone has a computer with a monster-size monitor. Luckily, you can resize tiles to fit your screen. Right-click a tile, point to *Resize*, and select the option that works best for you.

Change any setting with less work

You no longer have to hunt through windows and menus to tweak your mouse settings, desktop background, user account, or any other setting you can imagine. Old versions of Windows made you dig around in a Settings window or Control Panel to change these things. Now Windows will do the digging for you. Simply type a word about what you want to change in the Search box on your Taskbar.

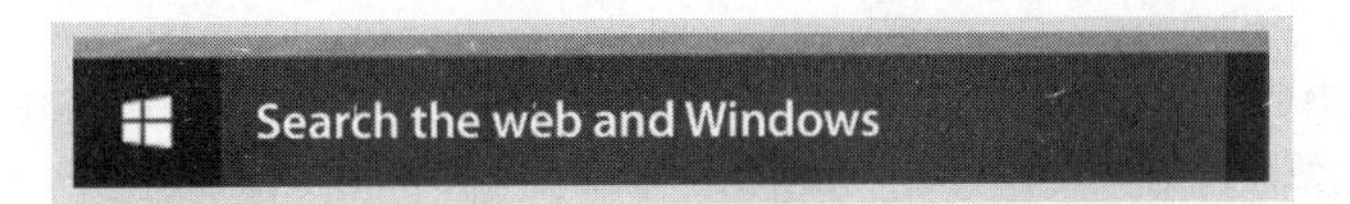

Say you're a lefty and want to reverse which button is the main click and which is the right-click on your mouse. Type "mouse" into the Search box and click the result that looks most promising. A window will open that takes you straight to mouse settings. Want to change your computer password? Type "password." How about your desktop background? Type "desktop background."

But what if you want to change something and you don't know what it's called? You can still poke around in the main *Settings* window and look for it yourself. Click ■ > Settings.

See social media messages on your desktop

Being a busybody has never been easier, what with Facebook, Twitter, and email. Staying on top of all those messages — that's the hard part. Or it used to be. Microsoft now corrals all of your social media messages and notifications into one place, right on your desktop. Here you'll also see reminders about events on your Calendar app, tips on using Windows 10, and important alerts like software updates. No more jumping between programs! It's all at your fingertips in the Action Center.

- Click ▤ in the far-right corner of your Taskbar to see new messages and alerts.

- Once you read them, you can delete each one by hovering over it with your cursor and clicking the X that appears. Or delete them all at once by clicking *Clear All* in the top-right corner of the Action Center.

You'll only see notifications for apps installed on your computer. For instance, if you want to see your Facebook messages in the Action Center, you need to download the Facebook app from the Microsoft Store. (Don't worry, it's free.) The same is true for Twitter, Instagram, and any app that doesn't automatically come with Windows 10.

What if your apps go crazy, pinging you with nonstop notifications? You can turn off those alerts.

1. Click *All Settings* at the bottom of the Action Center, then *System* in the window that pops open.
2. Click Notifications & actions > Show notifications from these apps.
3. Find the annoying app, click the *On* button beside it, and watch as it magically changes to *Off*.

Open a document right from your desktop

Nice work, butterfingers! You accidentally closed the letter you were writing, and now you have to root around your computer until you find it again. Save yourself the trouble. In many cases, you can reopen that file directly from your desktop.

1. First, pin your favorite programs to your Taskbar by clicking ⊞ > All Apps. Then right-click an app and choose *Pin to taskbar* from the pop-up menu. An icon for that app should appear on your Taskbar.

2. Right-click that icon to see the files you've worked on recently with that program. Simply click a file to open it.

These shortcuts, called Jump Lists, work for many programs, but not all. For instance, right-click the Taskbar icon for your word processor, and you may see all the files you opened in the last few days. Right-click your Web browser, however, and you might get nothing. It's a game of trial and error — Microsoft's favorite kind. Play with pinning programs to your Taskbar and right-clicking them to see what happens.

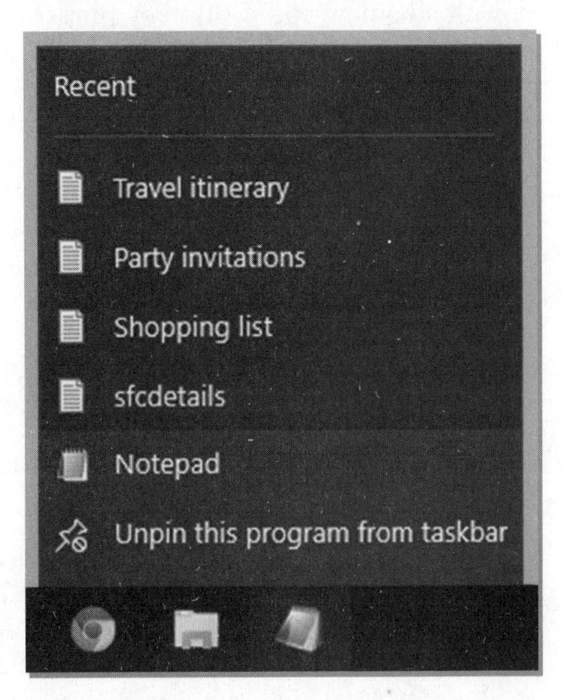

7 ways to get Help anywhere in Windows

No matter where you are in Windows 10, you're never far from a Help button. Although, to be honest, they ought to be called Help! buttons. Take a look at all the quick ways you can ask for assistance.

From the Start Menu. Need help with basics like how to connect to the Internet or how to change your user account? Click ⊞ > Get Started > Welcome to take a tour of Windows 10, or jump directly to a topic like Get Connected for help with a specific problem.

On the desktop. Click in the Search box on the lower-left corner of your Taskbar. You can type a whole question such as "How do I download a new app?" or a key phrase like "download apps." Click one of the search results to get an answer.

In any window. It's almost guaranteed that every window has a prominent Help button. You just have to know what they look like.

- Some windows have a blue question mark in their upper-right corner.
- Others have a Help menu in the upper-left corner next to the File, Edit, and View menus.

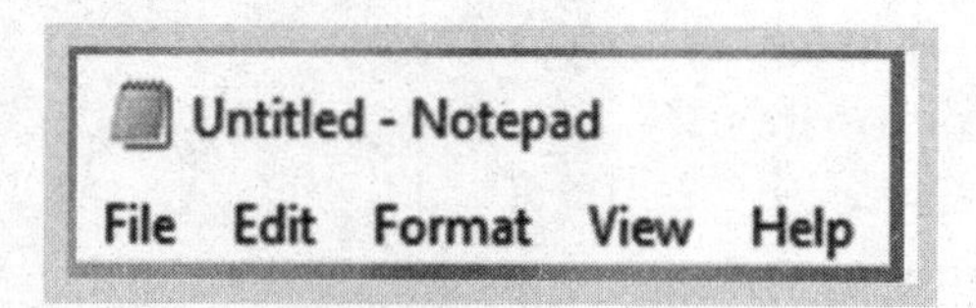

- Apps like Calendar and Mail have a Settings icon, like this one (⚙), which you can click to open Help.
- Other apps, including Google Chrome, hide their Help button behind an icon with three horizontal lines ≡.

Everywhere. Wherever you are in Windows, you're only an F1 key away from getting Help.

- Pressing F1 while on the desktop will pull up general help with Windows 10.
- Pressing F1 in a folder will open help topics for File Explorer.
- Pressing F1 in many programs will call up help for that program.

Turn on Tablet Mode when using a touchscreen

Some laptop computers have touch-sensitive screens, just like tablets and phones. The screen reacts when you tap it, which means you can click, double-click, and drag items with your fingers instead of your mouse. But tapping the tiny buttons isn't easy. That's why Windows 10 has Tablet Mode. The buttons and icons are bigger, and the Start Menu fills the entire screen — perfect for finger tapping.

You can switch between regular and Tablet Mode, depending on whether you're using a keyboard or a touch screen.

- To turn on Tablet Mode, click on the lower-right corner of your Taskbar to open the Action Center. Click *Tablet Mode* at the bottom of your window.
- To switch back to regular (aka desktop) mode, place your finger on the right edge of the screen and drag it toward the middle. This opens the Action Center. Tap *Tablet Mode* to turn it off.

Last but not least, a few hints on working in Tablet Mode. Apps fill up the entire screen, so it can be hard to switch back and forth between open programs. Swipe your finger from left to right across your screen to see all the apps you have open, and tap one to bring it full screen. Or tap the Back button on your Taskbar to go backward in an app or flip to a previous app. Drag an app to the bottom of your screen to close it.

Here's what that Windows key really does

The special Windows key can unlock tons of keyboard shortcuts and cut down the amount of clicking you do. Consider it turbocharged. Take a look at these handy uses on the following page, including some of the best new shortcuts of Windows 10.

Key combination	What it does
⊞	Open the Start Menu
⊞ + d	Show the desktop; press again to reopen all windows
⊞ + e	Open File Explorer
⊞ + i	Open Settings
⊞ + m	Minimize all open windows
⊞ + comma	Temporarily peek at the desktop
⊞ + t	Flip through your Taskbar programs
⊞ + Shift + m	Reopen all minimized windows
⊞ + Tab	Open Task View
⊞ + ↑	Maximize the window you're in
⊞ + ↓	Minimize the window you're in
⊞ + ←	Move the window you're in to the left side of the screen
⊞ + →	Move the window you're in to the right side of the screen
⊞ + Home	Minimize all windows except the one you're in; press again to bring back these windows
⊞ + Enter	Have Windows read aloud whatever is on the screen
⊞ + plus sign or minus sign	Zoom in (+) or out (-) using the Magnifier
⊞ + Esc	Close the Magnifier
⊞ + s	Open the Search box on your Taskbar
⊞ + l	Lock the computer or sign in using a different account
⊞ + a	Open the Action Center
⊞ + u	Open the Ease of Access Center

Skip the keyboard — dictate emails and documents

Anyone who takes the hunt-and-peck approach to typing will love the free dictation program built into Windows 10. You can dictate emails, Word documents, and almost anything else that normally requires a keyboard.

You'll need a good microphone — one sensitive enough to pick up your words and tune out background noise. The best mics for dictation plug into your computer through a USB port rather than a microphone jack.

Type "speech" in the Search box on your Taskbar and select *Windows Speech Recognition.* This opens the Speech Recognition Wizard, which walks you through the process of training your computer to understand what you say. It'll also teach you how to talk to your computer. Eventually, the Wizard will have you read a couple of sentences, so Windows can learn how you say certain words.

You can dictate into any Microsoft program. The trick is to open *Speech Recognition* before you open the other program. And while Microsoft programs may work with it, software made by other companies may not.

Enunciate but don't exaggerate. Feel your lips form each word, as if you're talking to someone who is trying to read your lips. Your computer will "hear" each word better and be less likely to make a mistake.

Speak the punctuation out loud. Say "period" at the end of a sentence and "new paragraph" to start a new paragraph. The same goes for commas, question marks, semicolons, and other punctuation.

Want to improve the computer's accuracy? Read some practice sentences. Type "speech" in the Search box on the Taskbar and choose *Speech Recognition* under *Settings.* Click *Train your computer to better understand you.* Then follow the instructions and read the sentences that Windows gives you.

Start with short notes. Don't begin by dictating that novel you've always wanted to write. Dictate simple things in the beginning, like short emails. Or read paragraphs from a book out loud, punctuation and all. This lets you focus on learning the different speech commands and correcting mistakes.

Correct mistakes with your voice, not your keyboard. By speaking the corrections, you basically train the dictation software and make it more accurate. Be sure to correct whole phrases, not just single words. This teaches the computer context.

For instance, maybe you said, "I'm famished," but Windows typed, "I'm finished." Simply say, "Correct I'm famished." Speech Recognition will jump to that phrase and show you several options. Say the number beside the right choice, and Windows will fix it. If none of the options are right, say, "Spell it" and do just that. Say "space" to put a space between two words when spelling them out.

Tell Windows to stop asking for your password

Bet you didn't know you belong to an exclusive club! Like a boys-only tree house, it requires a secret password every time you enter. It's your computer, and Windows 10 acts like a bouncer at the door. Walk away from your computer for a few minutes, and the screen goes dark as it slips into Sleep mode. Wake it up, and you're greeted with a "What's the password?" screen. Typing in your password 20 times a day can get old, fast.

You can change that, and it's easy. Type "password" into the Search box on your Taskbar and choose *Sign-in options.* This opens a Settings window, where you'll see a section labeled *Require sign-in.* Chances are, it's set on *When PC wakes up from sleep.* Click the drop-down arrow beside this phrase to open the menu, and select *Never.*

From now on, Windows will only ask for your password when you turn on your computer, not when it wakes up from Sleep mode.

'Print' letters and pictures without any paper

No printer? No problem! Windows 10's new Print to PDF feature lets you "print" articles, photos, and other files without a printer, ink, or paper. Instead of printing something on a piece of paper, the computer saves that article or image to your hard drive as a PDF file, so you can open it any time you want. You can also print the file for real at a later date.

Any program that lets you print things — a Web browser, word processor, or photo editing software, for instance — can do this. PDF files are perfect for sharing with friends because they are universal. That means anyone can open a PDF, no matter what kind of computer they use.

1. Click the Print option in the program you're in, as if you were going to print that file on paper. Often, that means clicking *File* in the upper left corner of the program and choosing *Print*, or pressing Ctrl+p on your keyboard.
2. This will open a Print box, where you can choose which printer you want to use. Choose *Microsoft Print to PDF*, and click the *Print* button at the bottom. (See next page.)

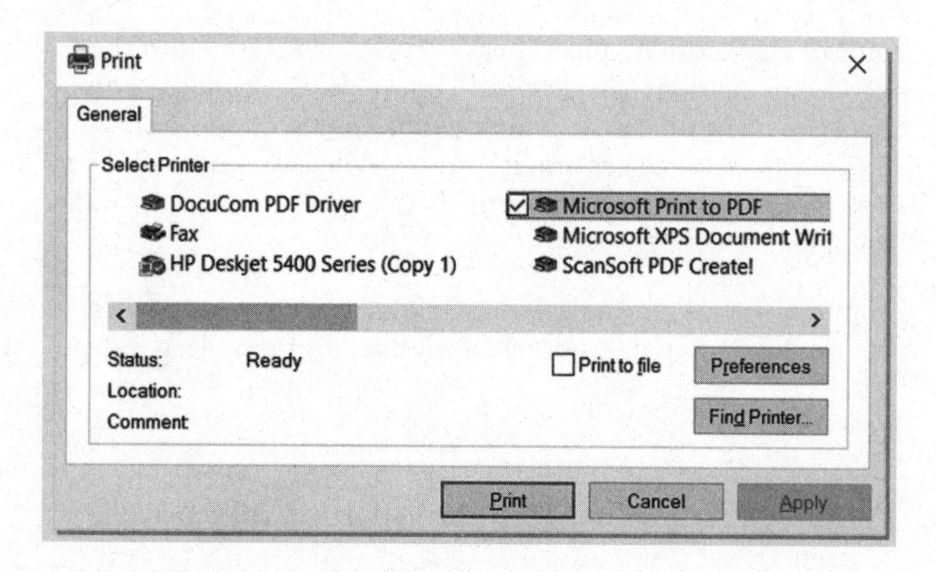

3. Name your file in the window that opens, and pick which folder you want to store it in. Click the *Save* button, and you're done!

Add more computer screens, for free

Ever wished you had two screens to hold all the open windows on your computer? Now you do. Windows 10 comes with Virtual Desktops, a feature that lets you create several desktops and flip between them while you work.

Stay organized. You can create a different desktop for each project you're currently working on, so that you don't end up with dozens of windows cluttering your screen. Are you organizing vacation photos, creating a scrapbook, and typing a letter? Create a desktop for each, and switch between them seamlessly without losing your place.

Share a computer with less aggravation. Say you and your spouse don't want the hassle of signing into the computer with different accounts, but you get in each other's way with so many windows cluttering the desktop. Put an end to that problem by creating a separate desktop for each of you.

- To create a new desktop, click ⧉ on your Taskbar, then click the phrase *+ New desktop* on the bottom-right corner of your screen. Or skip all the clicking and simply press ⊞+Ctrl+d.
- Once you've created at least two desktops, you can flip between them by clicking ⧉ again. All of your desktops will appear along the bottom of your screen. Click one to open it.
- To close a desktop, click ⧉, hover your cursor over one of the desktops along the bottom, and click the tiny X that appears above it.

Oops — did you forget to save the changes you made to a file before you closed that desktop? Don't worry. Any still-open windows will get plopped into another desktop.

Stay up to date and organized while you're on the go

You used to need a weekly planner to keep up with lunch dates, doctor appointments, and birthdays. Now you just need a phone or computer. With the Windows 10 Calendar app, appointments you add to your phone or tablet automatically show up on your computer's calendar — and vice-versa. Here's how to access and synchronize your computer and phone calendars wherever you are.

Using the calendar on your computer is simple. Click ⊞, then click the tile on the Start Menu with the date. That little square opens the Calendar app. If you don't see it, click *All apps* and scroll to *Calendar*.

Once open, you can add events and appointments by clicking directly on a date or *+ New event* on the left side of your Calendar window.

You may need to sign in with an email address the first time you open Calendar. If you want appointments and events to show up on your smartphone (Android or iPhone), then sign in with an Outlook, Hotmail, MSN, or Microsoft email address. These addresses work with Outlook, which you will need to install on your phone in order to see your Calendar appointments.

Get your appointments on the go. Use the app store on your smartphone or tablet to download the free Microsoft Outlook app. Log in to your Outlook, Hotmail, MSN, or Microsoft account when you first open the app. You may only see your email at first, but Outlook includes a calendar, too. Tap ≡ in the upper-left corner of the app window, then tap *Calendar.*

Give the Outlook app a few minutes to sync, or link up with, your computer's Calendar. From now on, any events you add to your calendar through a phone, tablet, or computer will appear almost immediately on your other devices.

Say 'hey' to your new personal assistant

The new Windows 10 has a built-in assistant named Cortana. No, she can't make coffee or bring you the paper. But she can put appointments on your calendar, tell you the weather forecast, and answer almost any question you have, whether it's about Windows or who won the World Series.

You'll have to jump through a lot of hoops when you fire up Cortana for the first time. She's nosier than a well-meaning friend asking about your blind date. She wants to know everything — including your location, your name, your email address, and your birth date.

Here's how to activate her and start putting her to work. Click in the Search box in the lower-left corner of your desktop. Then click the circle icon on the bottom left of the pane that opens. This pulls up Cortana. She'll explain what she needs and walk you through the steps. Keep these points in mind.

Let her track your location. Otherwise, she won't work. She'll ask politely, of course. "To help out, I need to use your location. You can give me permission to do that in Settings." Click the Settings button. This opens a window where you need to do two things.

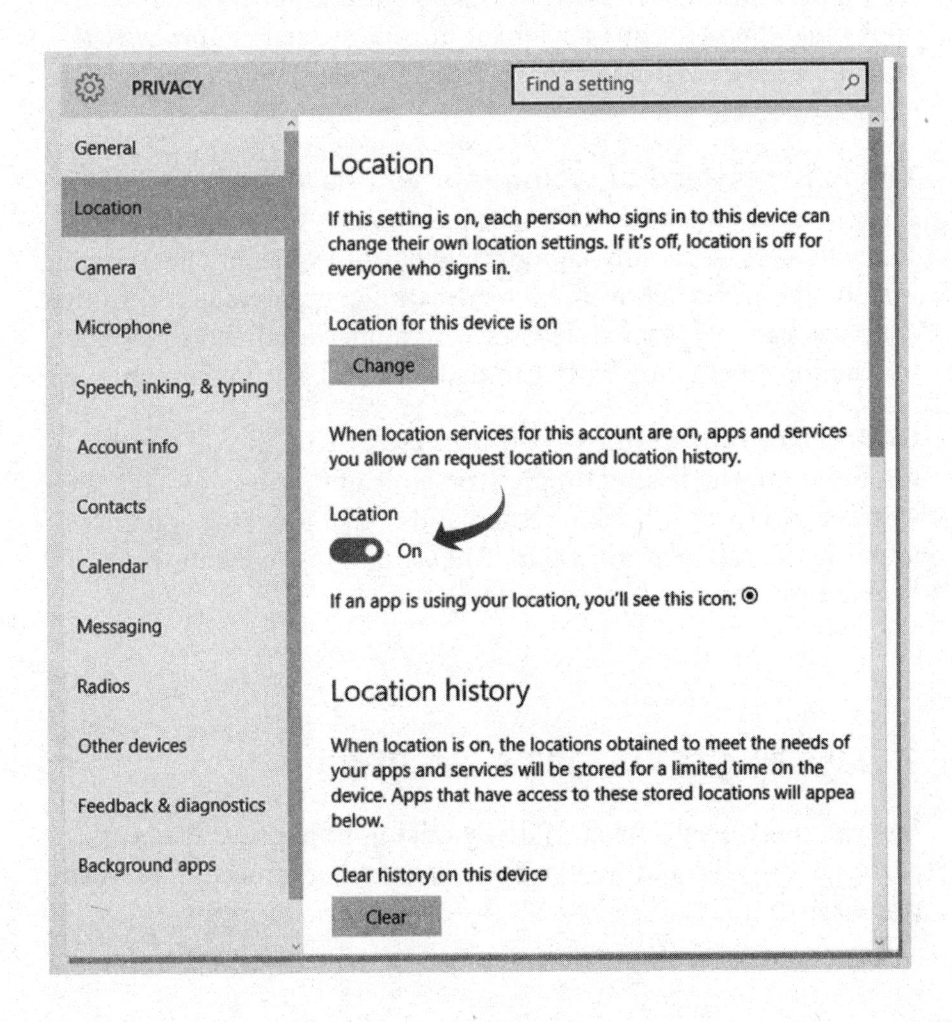

- Make sure this sentence reads, *Location for this device is on.* If it says *Off*, click the *Change* button. Click the word *Off* in the box to turn it to *On.*

- Now check the *Location* button below that. It should read *Location On.* If it says *Off*, click it to turn it to *On.*

Create a Microsoft account to use Cortana. It doesn't really make sense, but lots of things about Microsoft don't. Just roll with it. You may already have an account through Hotmail, Outlook, MSN, or Windows Messenger. If so, you're in luck! Just enter your info when Cortana asks for it during the startup process. If you don't have a Microsoft account, Cortana will give you a chance to create one.

Talk to her instead of typing — if you have a microphone. No, not the kind you belt "Sweet Caroline" into on karaoke night. Many devices, especially laptops, tablets, and smartphones, come with built-in microphones. Some desktop computers don't. In that case, you can buy a small headset with a microphone. Follow the instructions that come with it to make it work with your computer.

Or don't to talk to her if you don't want to. If you'd rather not deal with the hassle, simply type your questions ("What's the weather forecast today?") or commands ("Put a doctor appointment on my calendar for 3 p.m. August 6"). Just click in the Search box on your Taskbar and begin typing.

2 ways to get Cortana's attention

You can tell Cortana what to do by talking to her, just like a real personal assistant, but first you need to get her attention. Here are two ways to do it.

- Find the Search box on the left end of your Taskbar, and click the microphone icon. Cortana leaps to attention, hanging on your every word.
- Forget about clicking. Teach your computer to respond when you say, "Hey, Cortana." To do this, you'll need to change a setting in your computer. Click in the Search box on your Taskbar to open Cortana, and look at the icons on the left side. Click ▣ > Settings. Here you can flip lots of switches, including the one labeled *Hey, Cortana.* If this button says *Off*, just click it to *On*.

How to protect your privacy in Windows 10

Windows 10 has lots of shiny bells and whistles, but one new feature isn't much fun. Microsoft now uses the operating system to collect information about you, the same way Google and Facebook track your movements online.

It's all about learning what you like and who you're friends with, so these companies can sell you things and make more money. Fortunately, you can control a lot of what Windows (and Microsoft) know about you. Here are four tricks to do that.

Log in to your computer with a Local account, not a Microsoft account. When you set up a new computer, choose *Sign in without a Microsoft Account.* Do you already log in with a Microsoft account? You can change it to Local.

1. Type "account" into the Search box on your Taskbar, and click *Change your account picture or profile settings.*

2. Click *Sign in with a local account instead* in the center of the window that opens. Windows 10 will walk you through the process.

Be aware that using a Local account means you can't use Cortana, the Windows 10 personal assistant. You also won't be able to download apps from the Windows Store, or sync the Calendar app on your computer to your phone and tablet.

Turn off Cortana. This built-in personal assistant collects bits and pieces of information about you and radios it back to Microsoft. It's pretty harmless stuff, but you don't have to share it if you don't want to. Windows 10 comes with Cortana turned off, so if you've never used her, you don't need to do a thing. If, on the other hand, you set her up and took her for a spin, it's simple to disable her.

1. Click in the Search box on the Taskbar to open Cortana, then click ▣ > Settings > Cortana can give you suggestions, ideas, reminders, alerts and more.
2. Click the button from *On* to *Off*.

Keep online ads from stalking you. Do you get the feeling you're being watched when you surf the Internet? That's because you are. Microsoft, Google, and other companies are watching what you shop for, search for, read, and type. Then they show you ads for things they think you'll want to buy. Tell Microsoft to quit following you with this two-step process.

- Click ⊞ > Settings > Privacy > General. Look at the button beneath *Let apps use my advertising ID for experiences across apps.* If the button says *Off*, you're all set. If it's *On*, just click it to turn it off.

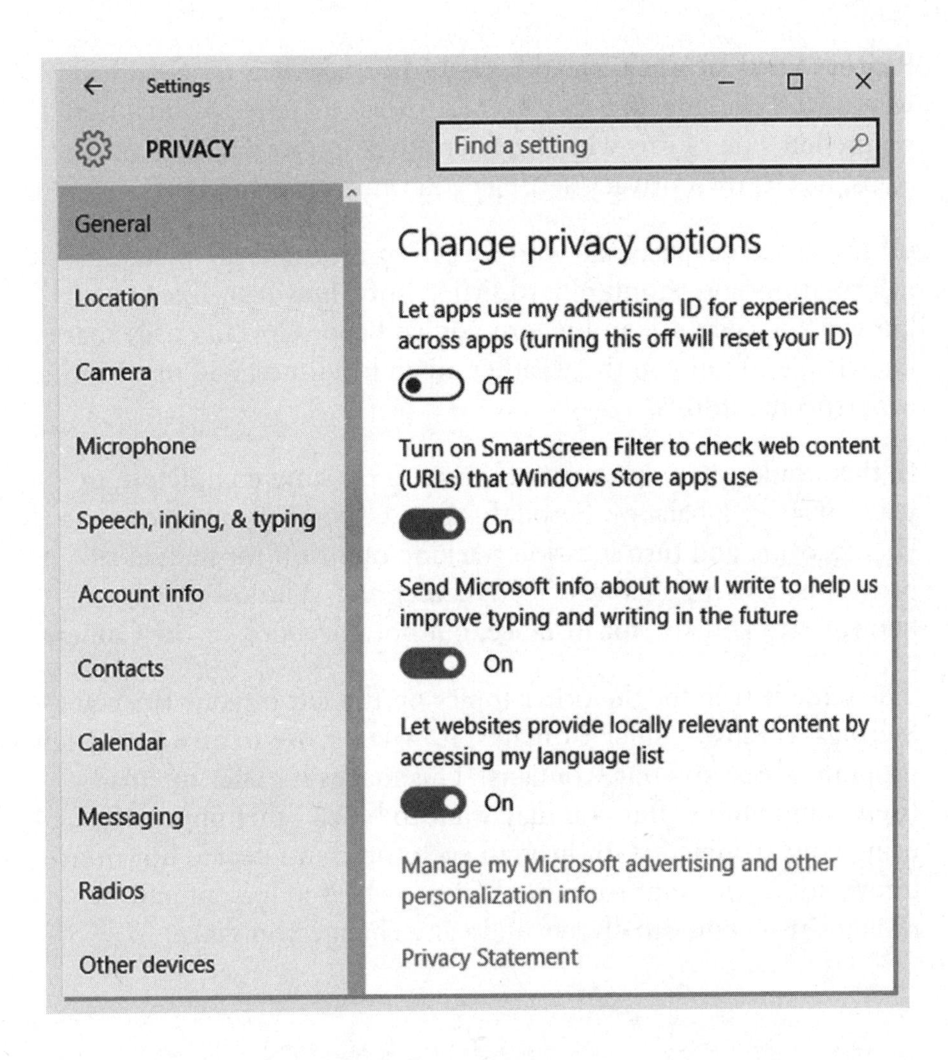

- In the same window, click *Manage my Microsoft advertising and other personalization info* at the bottom. This takes you to a website where you can tighten your privacy even more. On the right side, click the button beneath *Personalized ads in this browser* to turn it off. Now do the same with the button under *Personalized ads wherever I use my Microsoft account.*

Protect your privacy. Head back to the main Privacy Settings window by clicking ■ > Settings > Privacy > General. Each item on the left side of this window, from Location to Background Apps, has its own privacy settings you can turn on and off.

Of these, tech experts say your Location is one of the pieces of information you should guard. Microsoft, however, likes to know where you are and where you've been. Cortana uses that knowledge to tell you the weather, news headlines, and other zip code-specific tidbits.

In this window, you can turn off location tracking completely, or you can strike a balance. Scroll down to *Choose apps that can use your location*, and turn location tracking on or off for individual programs. You may want MSN Weather and Windows Maps to know where you are, for instance, but not Facebook or the Camera.

The same is true for the other topics on the left of your Privacy Settings window. Under Contacts, it makes sense to give your Mail program access to your Contacts. This makes it easier to email friends and family. But you may want to block other apps from using your Contacts. Go through each topic and decide how much information you want to share. Use your best judgment and remember — you can always undo any change you make.

No. 1 way to make your computer safer

There's one almost surefire way to protect your computer against hackers, viruses, spyware, and other baddies — sign in to your computer with a Standard account, not an Administrator account.

Being an Administrator gives you the power to install new software, delete software, and make changes deep inside Windows 10.

But if a hacker breaks into your computer or you accidentally download a virus while you're logged in as an Administrator, then that hacker or virus also has the ability to:

- delete good programs like your anti-virus software.
- install bad programs like identity-stealing software.
- change important settings deep inside Windows 10.

None of that can happen if you're logged in with a Standard account. Standard accounts let you surf the Web, use all of your programs, and download photos and other files from the Internet. But you can't install new software or change important settings. Since Standard accounts don't have the power to do any of those things, then bad guys don't either.

This gives you an extra layer of security against bad guys trying to trick you into downloading malicious software, or hack your computer and install software without you noticing. If you're using a Standard account, they can't. The minute they try, Windows 10 will ask you to type in your Administrator password. You'll know, of course, that you weren't trying to install any software, and you can click Cancel to shut down those crooks.

When you first set up your computer, you probably created an Administrator account. And that's probably the one you usually sign in with. Consider creating a Standard account for everyday use.

1. Click Windows > Settings.
2. Choose *Accounts* and Windows will show you details about the account you're currently signed into. If you see the word *Standard* beneath the account name, then congratulations! You're already signing in with a Standard account, and you

can stop here. If, however, it's labeled *Administrator*, then you'll need to create a Standard account, so keep going.

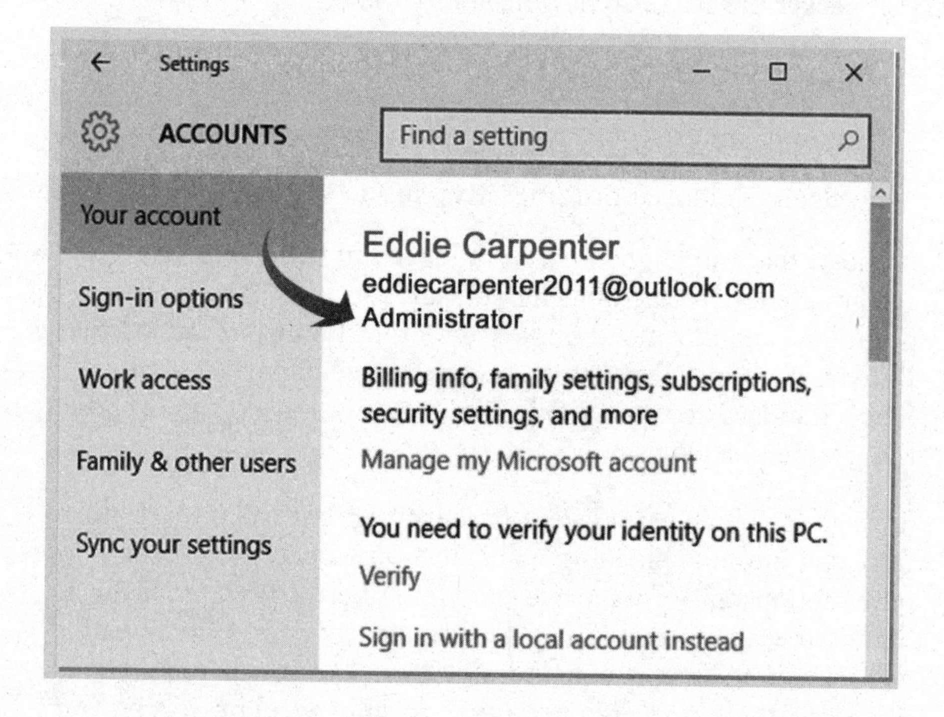

3. Click *Family & other users* on the left side of the window. Under *Other users*, click *Add someone else to this PC.*

4. Type in your email address. It can be the same one you use to sign into the Administrator account, or a different one.

   Windows wants you to use a Microsoft-based email address, such as Outlook or MSN. If you have one and choose to use it for the new account, simply type it into the email box, then click Next > Finish, and you're all set with your new account. But if you don't have a Microsoft email address, or if you don't want to use it, continue to the next steps.

5. In the window titled *How will this person sign in?* type the email address you want to use for the new account, and click *Next.*

6. You'll get an error message in red saying, *Looks like this isn't a Microsoft account.* That's OK. Simply click *sign up for a new one.*

7. This opens a new window titled *Let's create your account.* Click *Add a user without a Microsoft account* at the bottom of the window, then click *Next.*

8. In the next window, name your new account. It can be as simple as your first name followed by the word Standard. Type it into the *User name* box under *Who's going to use this PC?*

9. Pick a password and enter it beneath *Make it secure.* Give yourself a Password Hint to help jog your memory, just in case you forget it. Click *Next,* and you're done.

The next time you sign in to your computer, remember to choose the Standard account. When your computer boots up, both accounts will appear in the bottom-left corner of the sign-in page.

5 simple tips to speed up a slow computer

Is your computer as slow as molasses these days? Don't drop hundreds of dollars on a new machine. The one you have may be almost full and just in need of a quick tuneup.

Computers are like cars. They run smoothly 99 percent of the time, but occasionally they need new tires and an oil change. Consider the Disk Cleanup program your mechanic. It frees up space by emptying unnecessary files from your computer's memory banks, making it peppier and more nimble. Type "disk" in the Search box on your Taskbar, and click *Disk Cleanup.* Windows will run a quick check to see how many files you can safely delete, and how much space this will free up on your hard drive. Then, you'll see a window like this.

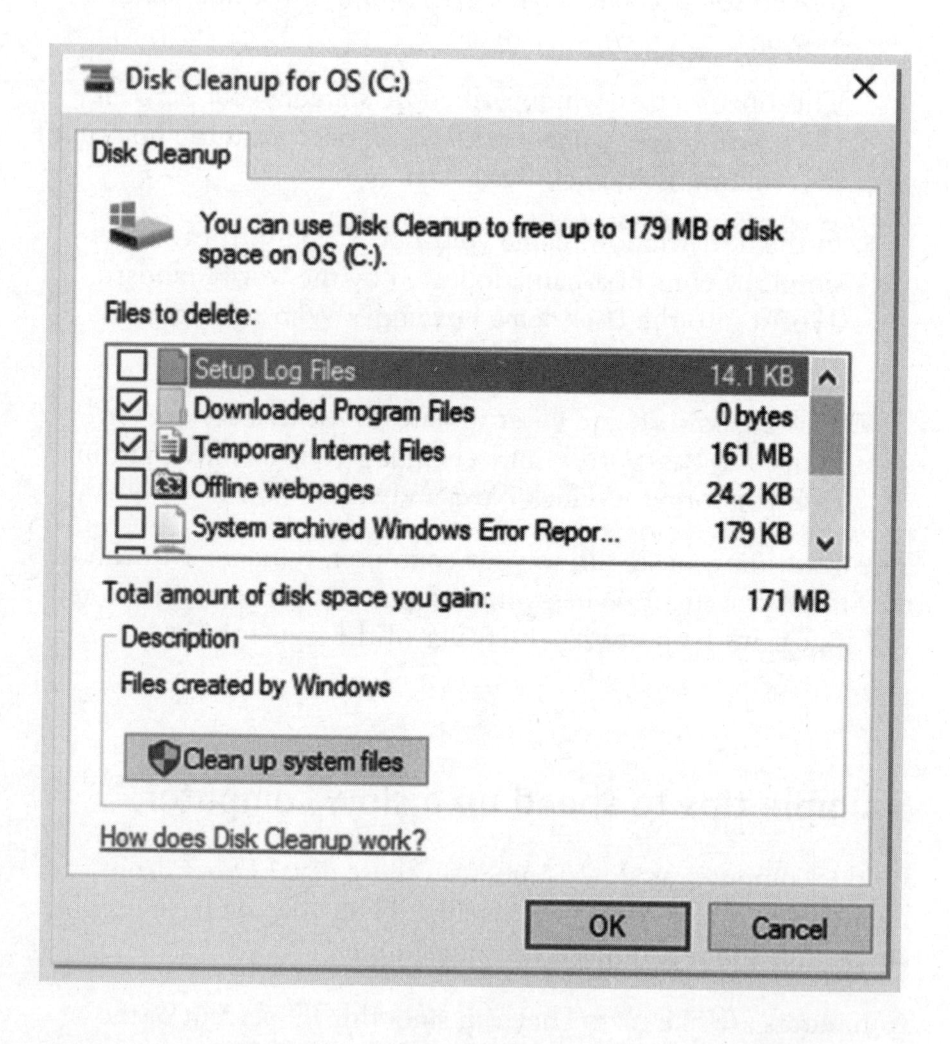

Now comes the fun part — dumping stuff. It's like cleaning out your attic, without having to haul bags of trash to the curb.

1. The idea of throwing out computer files can be a little nerve-wracking, especially if you don't know what they do. Click on any category in the list to see a summary about them in the space under *Description*. In general, you can safely delete these types.

    - Downloaded Program Files
    - Temporary Internet Files
    - Recycle Bin
    - Temporary Windows Installation Files
    - Temporary Files
    - System archived Windows Error Reports

2. Want to free up even more space? Click *Clean up system files* near the bottom of the Disk Cleanup window. Windows will dig deeper into its "attic" and come up with more files you can delete, including Setup Log Files and Previous Windows Installations. The latter can take up lots of space, especially if you upgraded to Windows 10 from an older operating system such as Windows 7 or 8. Check the boxes next to each type of file you want to delete, then click the *OK* button to clear them out.

3. After you click *Clean up system files*, you'll notice a *More Options* tab next to the *Disk Cleanup* tab at the top. *More Options* lets you delete programs you never use. Just click *Clean up...* and follow the instructions.

4. The same *More Options* tab also lets you delete old restore points from your hard drive. Your computer occasionally takes a snapshot of all its settings and saves those "pictures" to the hard drive. That way, if new software or devices start causing problems, Windows can rewind itself back to the last

time it worked smoothly, before the trouble started. That's great, but restore points can start to stack up. Clicking *Clean up...* under *System Restore and Shadow Copies* deletes all but the last restore point.

5. Disk Cleanup only deletes files created by Windows, the operating system. It doesn't touch leftover files created by other programs, like Firefox or Google Chrome. For that, you can download a free program such as CCleaner made by Piriform. It will clear out junk files left by non-Microsoft programs. You can get a basic version for free at *www.piriform.com/ccleaner*.

No. 1 thing to do before installing a new program

Stop! Before you install that new program on your computer, you should create a restore point. It could save you from disaster if the program short circuits your computer. A restore point allows you to "rewind" your machine to the moment before you installed the trouble-making program, so that everything runs smoothly again. Here's how to create one.

1. Type "restore" in the Search box on your Taskbar, and click *Create a restore point.* This opens the System Properties window.

2. Look at the *Available Drives* listed under *Protection Settings*, and click the drive labeled *(C:)*. To the right of the *(C:)* drive, you should see the word *On*. If you see the word *Off*, click *Configure*. Select *Turn on system protection* and *OK*.

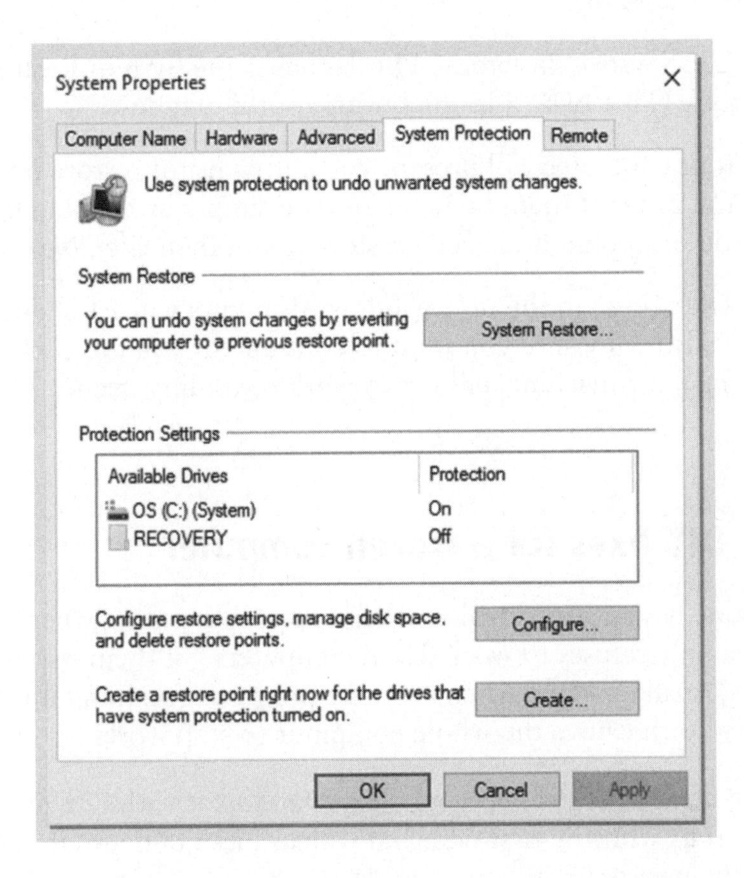

3. Back in the System Properties window, click *Create...* near the bottom. Pick a name for your restore point and type that into the box. Pick something descriptive that will jog your memory later, such as "Created before installing the Outlook Mail app." Then click *Create*, and you're done!

With any luck, you'll never need to use your restore point. But if the worst happens and a new program starts crashing your computer, here's how to rewind your computer back to a restore point.

1. Type "restore" into the Search box and click *Create a restore point.*

2. Click *System Restore....* This launches the System Restore tool. Click *Next >* in the bottom of this window.

3. Your computer will present you with a list of restore points. You can sort them by name or date simply by clicking the top of a column. Select a restore point, then click *Next >*.

4. Click *Finish* in the next window. Windows will ask if you're really, really sure you want to continue. If you do, click *Yes*, and your machine will get to work rewinding itself.

Easy DIY fixes for a frozen computer

Everyone has an occasional brain freeze — a moment where your mind simply refuses to work. Even computers get them. Computer freezes usually happen when a cranky program gets hung up, which in turn causes the whole computer to stop working.

Rather than panic, roll up your sleeves and fix it with one of these tricks. Start with the first one, and if that doesn't "thaw" it, move on to the next one.

- Walk away and get a cup of coffee. No, really. Give your computer a few minutes to think through whatever knot it's trying to untangle. Often it will work through the problem on its own.

- Press the Esc key twice on your keyboard. Sometimes that's enough to exit a stuck program.

- No luck? Press the Caps Lock key and watch for a tiny light on your keyboard to turn on. Then press Caps Lock again to see if that light turns off. If the keyboard responds, try these steps.

1. Press Ctrl+Alt+Delete on the keyboard and click *Task Manager* on the screen.

2. A window will open showing you all the programs currently running on your computer. Click the one you think is frozen and choose *End Task.*

3. Wait 10 to 20 seconds for that program to close, then see if your computer begins to work normally again. If not, open Task Manager again and close another program. Keep trying until you find the culprit.

- Still no luck? Restart your computer. Press Ctrl+Alt+Delete, and this time click the power icon in the lower-right corner of your screen. Choose *Restart* from the menu.

- Did anything happen when you pressed Ctrl+Alt+Delete? If not, then it's time to restart the hard way. Press and hold the machine's physical power button until it turns off. Wait 10 seconds or so, then press the power button again to turn it back on.

If none of these tricks work, then your computer may need professional help. Contact the manufacturer if it's still under warranty, or take it to a reputable computer repair store.

Rescue lost files fast with File History

One day the worst will happen and your computer will refuse to turn on. But you won't panic, because you've been using File History. You know that Windows backed up all of your files just yesterday, storing them safely on your external hard drive. Once you get a new computer, or get your old one fixed, you can get to work rescuing your files.

Plug the external hard drive into your computer, then type "restore your files" in the Search box on your Taskbar. Click *Restore your files with File History* in the search results. This will open a window showing you all the files Windows has copied and stashed away, grouped neatly into folders.

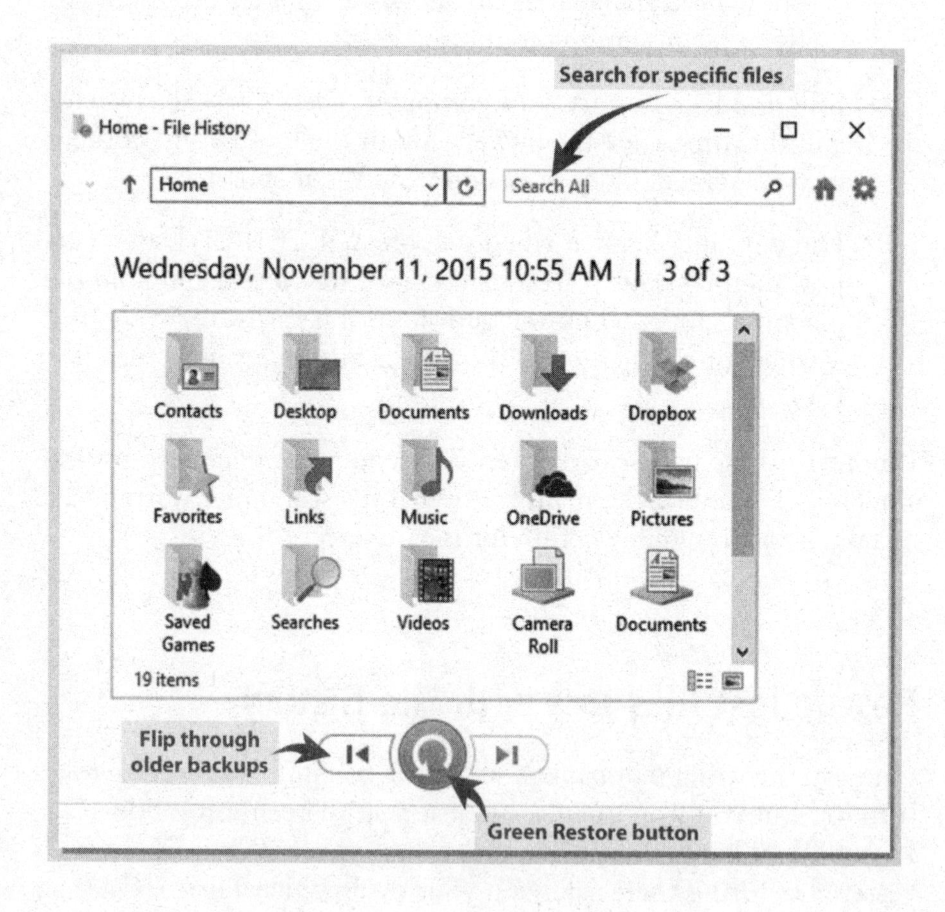

- If your entire computer crashes and you need to restore all of your files, simply click the green circle at the bottom of the window. Windows 10 will move everything back onto your computer's hard drive.
- If you lost only some of your files, you can restore just the missing ones. Click the folder that contains your lost files. You may need to click through several folders to dig down and find the one you need. Click to open it, so you can see what's inside. Then click the green circle at the bottom of the window.
- If you're just looking for a single file, type the name of it in the Search All box in the upper right of the File History window. Then click once more on the file's icon to open its contents. Now click the green circle to restore only that file.

Remember that Windows saves old backup copies, too, not just the copies it made an hour ago or yesterday. It has copies of files and folders from months, even years ago. So if you edit a letter, photo, or any other file and decide you don't like the changes, you can often retrieve the old version. Here are two ways to do it.

- Open File History as usual, and type the name of the file into the Search All box. Pay attention to the dates next to each file. Click the one with the date you're looking for, then click the green button at the bottom to restore it.
- You can also flip through old copies of the same file manually. Click through the folders in your File History window until you find the one you want. Now click the left-pointing arrow at the bottom of the window to flip through older copies. The time and date will change, telling you when that copy was made. When you find the version you want, click the green circle to restore it.

The computer will ask if you want to replace the current version of this file with the backup copy (*Replace the file in the destination*), or if you want to keep both. If you decide to keep both, choose the third option (*Compare info for both files*) and put a check mark next to each version you want to keep. The computer will add a number to the backup's name, such as "Dear John (2)" and store it in the same folder as the original "Dear John."

Do this, and you'll never need to double-click again

You can pretty much ditch double-clicking forever. All it takes is making one little change in Windows 10. That's welcome news to all the people whose fingers can't click fast enough for Windows to register it as a double-click, and for everyone whose carpal tunnel syndrome makes double-clicking painful.

Type "folder options" into the Search box on your Taskbar and select *File Explorer Options* in the results. This opens a window full of buttons and checkboxes where you can adjust what happens when you click on a file or folder.

By default, you have to click something to select it and double-click to open it. But click the button beside *Single-click to open an item (point to select)*, then *Apply* or *OK*, and that all changes.

Now every time you settle your mouse pointer over a file, folder, or other icon, you'll select it. And clicking it once will open it.

Word & Excel

Expert ways to turn work into play

The 1-click way to reformat text

Check out the first section of the Microsoft Word toolbar in your *Home* tab, and you'll see a paste icon, a set of scissors for removing text, and two pages that symbolize the copy feature. But what is that paintbrush, and what's it doing with your editing tools?

That odd little button is known as the Format Painter. Its purpose may not be immediately obvious, but once you learn how to use it, you'll turn to it again and again.

This icon allows you to make the text in one part of your document look like text from another part of your document — with the same font, size, color, and even indention. Simply highlight the text that has the flourishes you want, click the Format Painter icon, then highlight the text you wish to change. Just like that, your old content will have a whole new style.

If you want to change the formatting of more than one section of text, double-click the Format Painter icon to make it stay on. You can always exit Format Painter by pressing Esc or clicking the icon again.

7 basic keystrokes everyone should use

You use Microsoft Word and Excel all the time, but if you're not taking advantage of these easy shortcuts, you're missing out. Keyboard shortcuts can make everything you do on the computer faster and easier. You'll feel like a word-processing wizard in no time.

These same shortcuts work on Word and Excel for Mac, too. Just press ⌘ instead of Ctrl. Both keys can help you do basic tasks without all the clicking.

Ctrl+s	Save document
Ctrl+f	Find a word within document
Ctrl+c	Copy text
Ctrl+x	Cut text
Ctrl+v	Paste text
Ctrl+z	Undo typing
Ctrl+y	Repeat or redo typing

Paste plain text without the wild fonts

Larry Tesler. You may not know it, but he's your hero. As creator of the cut, copy, and paste commands, Tesler changed the future of word processing. Imagine where you'd be without them. (Hint: probably still typing up that document from last week, the one you pasted that article into.) The only flaw with copy and paste is that your computer copies and pastes not just the words you want, but all the funky fonts, sizes, and colors that appear in the original source. You could spend hours changing them all back to normal — but you don't have to. You can strip out all the special formatting with this easy trick.

- On a Mac, right-click (or press Control and click) where you want to paste the text. Select *Paste Special* from the drop-down menu, then choose an option from the dialog box.

- In Windows, simply right-click and select one of the paste options. *Keep Source Formatting* retains the text size, style, and color from the original source. *Merge Formatting* matches the text to your current font and size but keeps any bold and italic markings from the original. *Keep Text Only* strips away all the original formatting, matching the text to your current style, size, and color.

These same options also live under the *Home* tab on both PCs and Macs. Just click the arrow beneath *Paste*. Here, you can also select *Paste Special* and choose one of the more advanced options. Skip all the clicking by pressing the shortcut keys Ctrl+Alt+v (Windows) or Control+⌘+v (Mac).

Microsoft Office isn't the only program with special paste options. You can find a similar feature in applications such as Apple Mail, Pages, and Numbers as well as Adobe InDesign and OpenOffice.

Where to find special characters and symbols

Your keyboard has plenty of keys, including a few that come in handy when you hit your thumb with a hammer, like @, #, and !. But start searching for ö, ©, or ≤, and you're out of luck. Seldom-used symbols like these are hidden from plain view. Well, now you can stop searching for them. Autocorrect already replaces certain text as you type — for instance, (C) changes to © and (tm) to ™. Now, you can add your own replacements.

- In Windows, add your most frequently used symbols by clicking File > Options > Proofing > AutoCorrect Options and selecting the *AutoCorrect* tab.
- On a Mac, click Word (or Excel) > Preferences > AutoCorrect, and then click the *AutoCorrect* tab.

There, you'll see a table of character combinations and what symbols they turn into. Make a cheat sheet for yourself by writing down the character combos for symbols that you use often. Of course, sometimes you'll want to type a symbol that isn't in the table. No problem. In Word, just click the *Insert* tab, then the *Symbols* (Windows) or *Advanced Symbol* (Mac) button.

Your Mac gives you an even easier way to insert symbols in many different programs. Learn how on page 58.

2 simple ways to check grammar and spelling

It's your word processor's way of saying, "Hey, I think you might have typed this wrong." Affectionately called the "squiggly mark," the Spelling & Grammar checker's underline has spared lots of people from embarrassing moments and probably saved a few jobs.

It's easy to understand. The red squiggly signals a spelling error; the green (Mac) or blue (Windows) squiggly calls your grammar into question. Here are two ways to get the most out of this iconic feature.

- In Windows, right-click the underlined word or phrase and select the correct option from the drop-down menu. On a Mac, press Control and click the word. Select *Ignore All* to reject Word's suggestions or *Add to Dictionary* if you want the checker to learn that word.
- The *Spelling & Grammar* check button is under the *Review* tab. Click this button if you want to go through each error one by one. It will provide suggestions and make sure you don't accidentally miss a mistake.

Split long documents in two for easier viewing

Long documents and large spreadsheets are a pain, with all the back-and-forth, up-and-down scrolling they require. You may be typing at the end but constantly having to refer to info at the beginning. Wouldn't it be nice to split that document into two pieces, side-by-side? You can, thanks to Split view.

You may not have noticed this feature, but it's been around awhile. Now it's easier to find. Under the *View* tab, click *Split.* This feature splits the screen into two halves that you can scroll through independently. Compare one part of a document to another, or refer back to something brilliant you wrote without the frustration of searching.

In fact, this feature is so handy it even has its own shortcut. In Windows, press Alt+Ctrl+s. On your Mac, press Option+⌘+s.

Use Tell Me to find nifty features fast

You probably don't have the recall skills of Ron White, two-time national memory champion. So when it comes to navigating Office, it's hard to remember all the options, especially the ones Microsoft has moved since the last version of Word. If you have a PC, though, you're in luck. The Tell Me bar will help you do what you need without all the memorizing and searching.

You'll find this feature on the right end of the Menu bar, beside the light bulb. It's the search field that says *Tell me what you want to do.* Do exactly that by typing your question into this box, and Tell Me will give you several choices based on your query.

Remember Clippy, that pesky paperclip that popped up in old versions of Office, always offering unsolicited advice? Yeah, Tell Me isn't pushy like that. What it is, is way more advanced. It can understand your most human-sounding commands, such as "Put a bullet here," "Make that italics," "Change how this table looks," and "Share my document." Jump to the Tell Me search field faster by pressing Alt+q.

Share programs, save money with Microsoft Office Home

Just about everyone uses Word and other Microsoft Office programs, but they're so expensive! If you want Office without sacrificing an arm and a leg, try the 365 Home subscription. You'll get more bang for your buck with this deal. The Microsoft Office 365 Home subscription has a few advantages over other plans.

- The new Office 2016 works on many different devices, not just computers. That means each person can also download Office to their phone or tablet.

- Up to five people can share it, because you're allowed to download it on as many as five computers. That may be the best deal out there, considering the Home plan costs only $9.99 a month (or $99.99 per year), while a one-computer subscription runs $6.99 a month (or $69.99 per year).
- You can try it free for one month.
- Each person can store up to 1 terrabyte (TB) of data (documents, pictures, music, and more) in the cloud for free using Microsoft's OneDrive service.

Microsoft Office comes with lots of bells and whistles, like templates for creating budgets, party invitations, and flyers — maybe more than you need. Some bare-bones word processing programs, such as Google Docs, are free. See page 252 to learn more about the free document and spreadsheet programs available through Google Drive.

Rescue files you closed without saving

Oh no. You've opened a document to continue your work from yesterday, when you notice that some of the changes you made — important changes — are no longer there! It's as if you forgot to save them. Now what do you do? You can't possibly remember every word you wrote yesterday.

You don't have to. Just take a look at your revision history, and you'll be back on track in no time.

- In Windows, any documents with unsaved changes appear under File > Info. Check the *Manage Document* section (called *Manage Workbook* in Excel). You can open these unsaved versions and compare them to the one you currently have. When you find the right one, click *Restore* to begin working in it.

- On a Mac, go to File > Restore and choose either *View Version History Online* (to view older versions saved in OneDrive) or *Restore to Last Opened.*

Select text 1 word, line, or paragraph at a time

Have you ever tried to highlight a section of text by clicking and holding as you scroll down the page, only to lose your text and have to do it all over again? Stop dragging your cursor all over the place. There's an easier way to grab exactly what you need. You can select:

- **one word** by double-clicking it.
- **more than one word** by clicking at the beginning of a word, holding down the mouse button, and dragging your cursor across the other words.
- **an entire line** in a Word document by clicking next to the line in the left margin.
- **a whole paragraph** in a document by triple-clicking a word in that paragraph or double-clicking in the left margin beside the paragraph.
- **a lot of text** by double-clicking the first word and pressing shift while clicking the last.
- **everything** in a document by pressing Ctrl+a (Windows) or ⌘+a (Mac).

Change ALL CAPS to lowercase in less than 3 moves

You find an article with great tech advice, and you copy the most important parts to a document so you can keep it forever. But wait — THE FIRST LINE OF EVERY PARAGRAPH IS IN ALL CAPS. How frustrating. Before you give up on the whole thing, try this tip to turn capitals into lowercase. First, highlight the troublesome text. Then:

- in Windows, press Shift+F3.
- on a Mac, press Fn+Shift+F3.

Each time you press this key combination, the program cycles through a different option, from all caps, to all lowercase, to capitalizing the first word in each sentence. You don't have to remember a shortcut, though. Just click the double-a Change Case icon under the *Home* tab to see these options and more. Aa▾

Why words can disappear as you type — and how to stop it

It's every novelist's nightmare. The muse strikes, and you're typing fast, fingers flying over the keyboard as you frantically try to capture the flow of words in your head. Suddenly, you look up at the screen and discover — gasp! — that you have typed over every word you've written in the last three days. Now that's a real horror film! And if it's ever happened to you, you know firsthand the panic of Overstrike mode.

How did it happen? More importantly, how can you keep it from happening again? If you're a PC user, chances are you accidentally pressed the Insert key on your keyboard when you meant to press Delete. Overstrike (or overtype) mode does exactly what it says — it writes over the existing words in a document, and it can make you think you're losing your mind.

Luckily, all you have to do is press the Insert key again to turn it off and return to regular mode. If this happens to you a lot, though, you can strip the Insert key of its power. Open Word and click File > Options > Advanced, then uncheck the box beside *Use the Insert key to control overtype mode.*

On the other hand, you may think this function is far-out fantastic. But what do you do if you're a Mac user? Your keyboard doesn't have an Insert key. Instead, open Microsoft Word, click *Word* in the upper-left corner, then click Preferences > Edit. Click to check the box beside *Replace existing text as you type (Overtype mode).*

Streamline group work by sharing documents

Give me a T! Give me an E! Give me an A! Give me an M! What does that spell? TEAM! Whether you're completing a work project with colleagues or creating a digital photo album with your mother-in-law, the truth is group projects aren't all that fun. Word 2016 has people singing a different tune, though.

That's because you can share one document with your entire team. Co-authoring allows the whole gang to work on a document at the same time and to comment on each other's work — without being in the same room. Working with your mother-in-law just got a lot more pleasant.

Open the file and click the *Share* icon in the top-right corner of the window. This opens the Share pane. Office will prompt you to save a copy in the cloud if you haven't already. (You get free cloud storage with OneDrive when you sign up for a Microsoft account. Learn about this and other cloud storage options on page 261.) Once it's in the cloud, you have several ways to share it with your team.

In Windows, open the Share pane and:

- type people's email addresses into the *Invite people* field, or click the *Address Book* to the right of it and add their emails directly from your contact list. You can even add a little message to get them pumped up about working with you. Next, decide whether you want to let your team members make changes to the file or just view it. Select *Can edit* or *Can view* from the drop-down menu, then click *Share*.

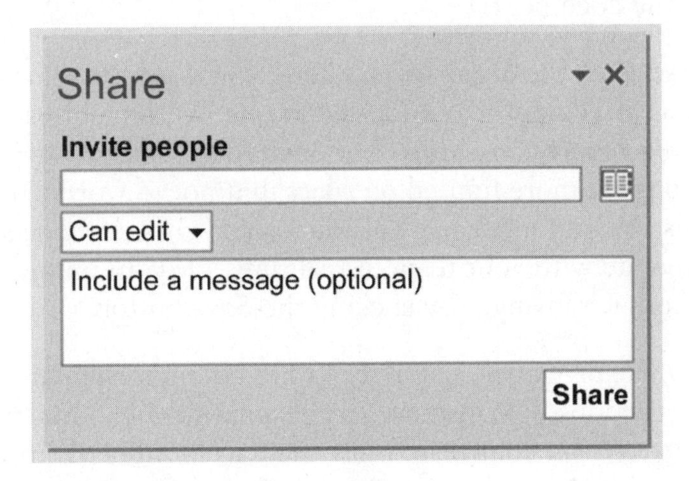

- click *Get a sharing link* at the bottom of the Share pane. Again, decide whether you want other people to be able to edit the file or simply view it. Then click the appropriate button under either *Edit link* or *View-only link*. You can copy and past the link that appears into an email or even a Facebook post.

- click *Send as attachment* at the bottom of the Share pane. Use this option to email your team either a copy or a PDF of the file.

When someone opens the file and starts making changes, a box may pop up on your screen asking if you want Word to save these changes as they happen. You can always change your mind later by clicking File > Options > General.

Look for *Real-time collaboration options* at the bottom of the window, click the drop-down menu below it, and choose when you want to share your changes. You can also check the box next to *Show names on presence flags*. This enables you to see who changed what in the document.

On a Mac, click the *Share* icon. There, you'll see the same sharing options as in Windows, but located in one, convenient list. Choose from *Invite People, Copy Link*, and *Send Attachment*. Real-time collaborations are more limited on Macs, but not to worry. You can always use Word Online, part of Microsoft's Office Online software, to collaborate with your team in real time. Or you can manually sync everyone's changes by clicking the *Save* button.

Other Office programs, like Excel, offer co-authoring but with more restrictions. For instance, if you want to collaborate in real time, everyone on your team must use Excel Online. You can do the same with Google's suite of free software, too. Learn how to share and edit all kinds of files in real time, for free, with Google Drive on page 252.

Choose how and when Word corrects your text

"No, I didn't want to do that!" Usually, Word's AutoCorrect and AutoFormat features save you time by fixing your spelling and grammar, neatly numbering your lists, whipping up tables, and transforming web addresses into active links in your document. But when Word constantly "corrects" things as you type, it can get pretty annoying. Luckily, there's an easy fix.

Every time Word changes something you type or paste, a lightening-bolt symbol pops up.

Clicking the arrow beside it opens a menu where you can undo that change or turn off autoformatting for good.

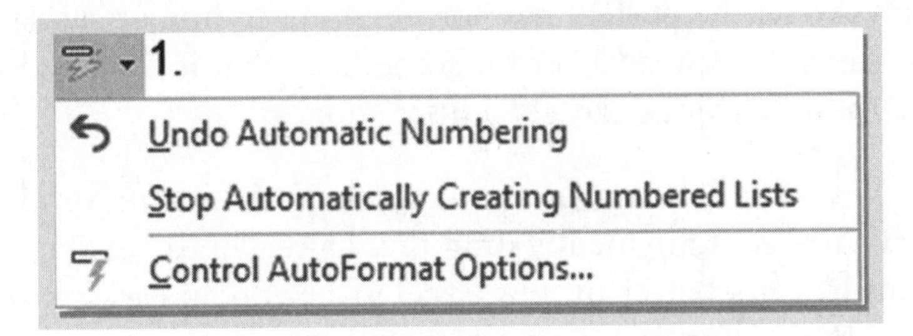

You can also undo autoformatting without ever touching your mouse. Press Ctrl+z (Windows) or ⌘+z (Mac) immediately after Word changes something.

AutoFormat comes in especially handy when you create a numbered or bulleted list. Unfortunately, it's not smart enough to know when you've reached the end of your list. You think you're finished, press Enter, and then watch helplessly as Word automatically begins the next line with another number. Here's how to take control.

- To keep the indentation but remove the bullet or number, press Backspace.

- To remove the indentation as well as the number or bullet, press Enter again.

It may not always feel like it, but remember — you have complete control of your AutoFormat settings. In Windows, adjust them by clicking File > Options > Proofing > AutoCorrect Options > AutoFormat As You Type. On a Mac, click Word > Preferences > AutoCorrect > AutoFormat As You Type.

Instantly adjust rows and columns in Excel

The computer nerds who developed Excel fashioned the columns and rows to be the perfect size for plugging in numerical data. If you're using the spreadsheet for something that involves a lot of words, though, things can get a little snug.

What do you do when your text is too long to sit comfortably in the cell? You could manually drag the divider lines between columns to get a better fit, but Excel gives you an easier way. Get ready to leave your dragging days behind you.

You can also resize more than one row or column at a time. Just highlight all the ones you want to adjust, then double-click one of the divider lines.

Wrap text to keep content from getting cut off

If you're not an Excel frequent-flier, your experience with spreadsheets may be filled with frustration. Why does your text overlap into the next column or get cut off? Auto-adjusting the column will make it wider, but if you're filling the cell with a paragraph, you don't want the cell to stretch clear across the country. Here's what you do.

Select the problem cells or column. Under the *Home* tab, click *Wrap Text*. This feature will wrap the text so that it fits in the cell's current width. Problem solved.

Wrap Text

3 ways to edit a cell in a jiffy

You know how Excel works. It's the master of manipulating cells to create something epic. Sometimes, though, the most basic things about Excel are a mystery. Take typing stuff into cells. In an empty cell, all you need to do is click it and start typing. That's not the case if a cell already has text in it. Here's a primer on when to click, when to double-click, and when to press F2.

- To overwrite the text in a cell, click it and start typing.
- To edit the text in a cell, double-click the cell to put your cursor in it, then start typing. The cursor will appear exactly where you double-clicked.
- To put your cursor at the end of the text in a cell, click the cell once, then press F2 (Windows) or Control+u (Mac).

Copy and paste without all the clicking

They may have started out with glue, paper, and a pair of scissors, but copying and pasting have gone digital. These tools live in lots of electronics, from the most basic computers to the fanciest smartphones. That includes spreadsheets. Excel offers two ways to copy and paste text from one cell into another.

Copy with a keystroke. This handy shortcut duplicates the contents of one cell into the cell below it. Click the cell below the one you want to copy and press Ctrl + quotation mark (") in Windows, or Control + quotation mark (") on a Mac. For instance, if you want to copy the text in cell A2 and paste it into A3, click A3 and press the key combination.

Fill with a flick of your mouse. The shortcut above is great for copying a number or formula. However, if you want to paste it into more than one cell at a time or want the copied formula to reformat, this will really blow your mind.

When you select one or more cells in Excel, a small square appears in the bottom-right corner of the selection. You may think it's just for decoration, but it's actually called the Fill Handle, and it has a nifty purpose — to copy data into neighboring cells.

Unlike the shortcut above, the Fill Handle automatically tweaks formulas to match the cells you paste them into. So if the formula in C1 referred to cells A1 and B1, then copying it into C2 will change it to A2 and B2.

This handy handle can also copy and paste into more than one cell at a time, another bonus. First, select all the cells you want to copy. Then click the Fill Handle and drag it over the cells you're pasting into, whether they're above, beside, or below. Drop the Handle and presto! The cells fill in like magic.

After dropping the Fill Handle, you can click the icon that pops up to change exactly how Excel fills in the cells. (See example on next page.)

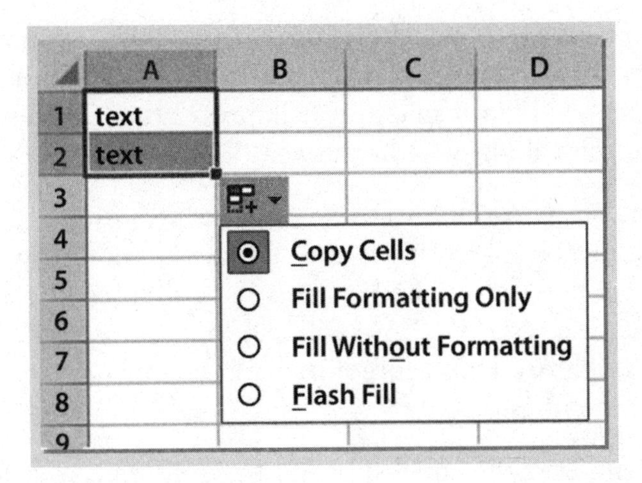

Easy guide to building Excel formulas

Excel isn't just for creating tables. It can do amazing things, from simple actions like adding and subtracting to more complicated feats like calculating a mortgage loan, solving engineering problems, and organizing time management charts.

It all comes down to formulas. Knowing how to use formulas in Excel is mighty useful. Bonus — you'll look like a math whiz.

When you need to use a formula, start with an empty cell and an equal sign (=). From there, the possibilities are endless. Here are a few to get you started.

- The capacity to reference other cells is what makes Excel a spreadsheet powerhouse. For instance, if you want to add the number in cell A1 to the number in cell B1, you can type "=SUM(A1,B1)" into any empty cell, press Enter, and see the answer. Cell referencing is the foundation for complex spreadsheets.

- No need for a calculator! Just type numbers and math operators (like plus or minus signs) to get your answer. For example, type "=1+1" into an empty cell, press Enter (or Return) and the cell will display the answer, "2."
- Use preset formulas. A function like =SUM(1,1) will add the numbers in parentheses and display the total. You can use SUM functions to add up cells in your spreadsheet, too. For instance, =SUM(A2:A6) will add the numbers in cells A2 through A6. Don't forget to enclose your function in parentheses.

When you insert the equal sign and begin to type your formula, Excel automatically gives you suggestions. But if you don't find what you're looking for (or just can't remember), there's an easier way. Go to the *Formula* tab and click *Insert Function*. The *Formula Builder* will give you a list of functions and their definitions, so you'll know how to use them.

Fastest way to write a formula

You can probably think of more exciting things to do besides typing formulas into Excel, especially when it comes to referencing other cells. You look up the column letter. You look up the row number. You type it into your formula. You sigh, insert a comma, and go hunting for the next one.

Stop right there! When writing a formula, you don't need to type out every single cell. Just click them to fill in the formula automatically. Here's the play-by-play. After you type the first parenthesis, click the first cell you want to include in the formula. Enter a comma, click the next cell, and so on. (See example.) Once you've included all the cells you want, type the end parenthesis. Boom. Now you can get back to doing things you actually want to do.

SUM | × ✓ *fx* | =SUM(C3,D2,D3,E4,F3)

	C	D	E	F	G
1	F3)				
2	2	29	19	6	
3	16	8	11	4	
4	0	31	18	23	

Super-secret way to simplify spreadsheets

Formulas are Excel's big gift to the world of spreadsheets. They can automate all sorts of calculations — but they can also be a bur under your saddle if you don't know how to use them.

Say you wrote a sweet little formula that totaled your expenses from Column B. Now you want to add up the expenses in Column C. No need to write a brand new formula!

Simply copy and paste the one from Column B into an empty cell at the bottom of Column C. The cells in the formula will automatically change to reflect the new column. For instance, =SUM(B2:B10) will become =SUM(C2:C10). Every time you copy a formula from one cell and paste it into another, Excel will tweak it to reflect the new location.

But what if you don't want that formula to change? What if you want to paste it into a new cell, but you still want it to show the total of B2 through B10? Excel can do that, too. It just takes a little $ — the dollar symbol, that is. The tip is totally free. Put a dollar sign ($) in front of each letter and number in the formula. So instead of typing "=SUM(B2:B10)," type "=SUM($B$2:$B$10)."

Want to get crazy? You can force a formula to keep the rows constant but change columns when you paste it into a new cell. For instance, type "=SUM(B$2:B$10)" into cell B11, then copy and paste it into cell C15. What happens? The formula still adds up rows 2 through 10 — but of Column C.

You can cycle through all these formula options fast with a lot less thought. Highlight the formula then press F4 (Windows) or ⌘+t (Mac).

Have fun with hidden spreadsheets

Picture this. You're scrolling down an Excel sheet, looking for those stats your friend promised, but they are nowhere to be found. What gives? Chances are, your workbook has more than one spreadsheet. You just have to know where to look.

Open Excel file and scan the bottom of the spreadsheet for tabs labeled "Sheet1," "Sheet2," and so on. These tabs may hide additional spreadsheets that are all part of the main file. Click these to flip through them. You can add more sheets by clicking the + icon at the end. You can even jump between those sneaky sheets by pressing Ctrl+Page Up or Ctrl+Page Down (Windows).

Working in multiple spreadsheets is easier (and more fun) when you customize them. Right-click a *Sheet* tab and select *Rename* to label it whatever you like. Right-click again and choose *Tab Color* to add a splash of color or *Move or Copy* to rearrange them.

Email

Take control of your Inbox

See all your email accounts in one convenient place

Stop signing in to all of your email accounts separately! Those are precious seconds of life that you'll never get back. If you have multiple email addresses, the Windows 10 and Apple Mail apps can simplify things by syncing all your accounts, so that you see all of your emails in one place.

- On a computer running Windows 10, press ⊞ to open the Start Menu. Then click the Mail app > Accounts > + Add account. Pick the type of account you want to add, such as Google or Yahoo! Mail. Fill in your username and password for that account and click *Sign in.* Microsoft will add it to the left-hand panel of your Mail window. Simply click the different accounts on the left to see the emails in each one.

- On an Apple computer, open your Mail app. Then click *Mail* in the Menu bar along the top of your screen. Select *Add*

Account, pick the type of account you want to add, and enter its username and password. Mail will remember this info, and all of your emails will appear in the same window from now on.

Clear Inbox clutter with Gmail filters

Gotta love Gmail. Not only is it the most popular email service out there, it also comes equipped with an amazing filtering system that will automatically organize your email Inbox for easy use.

First, purge some of the messages you never read. An easy way to slash the number of emails you receive is to unsubscribe from social media alerts, coupon deals, and newsletters. Once you've done that, you're ready to filter the barrage of emails you still get.

1. Go to the search box in Gmail and enter your search terms. Say you want to limit emails that contain the words "new deals." Type that phrase into the search box.

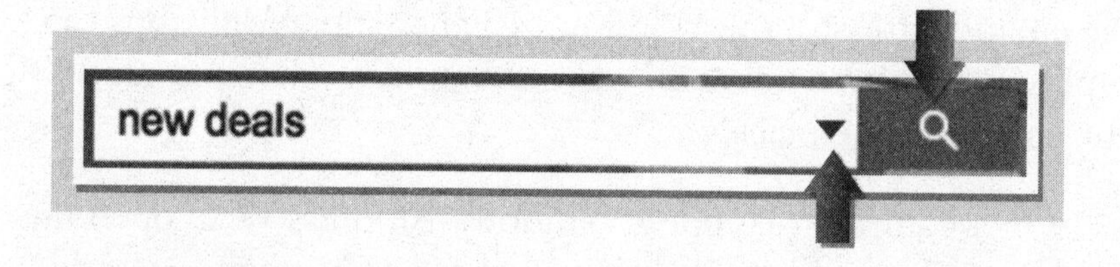

2. Click the drop-down arrow in the search box to further define your criteria. You can type the name of a specific sender, add key words, and even specify a message size or time frame.

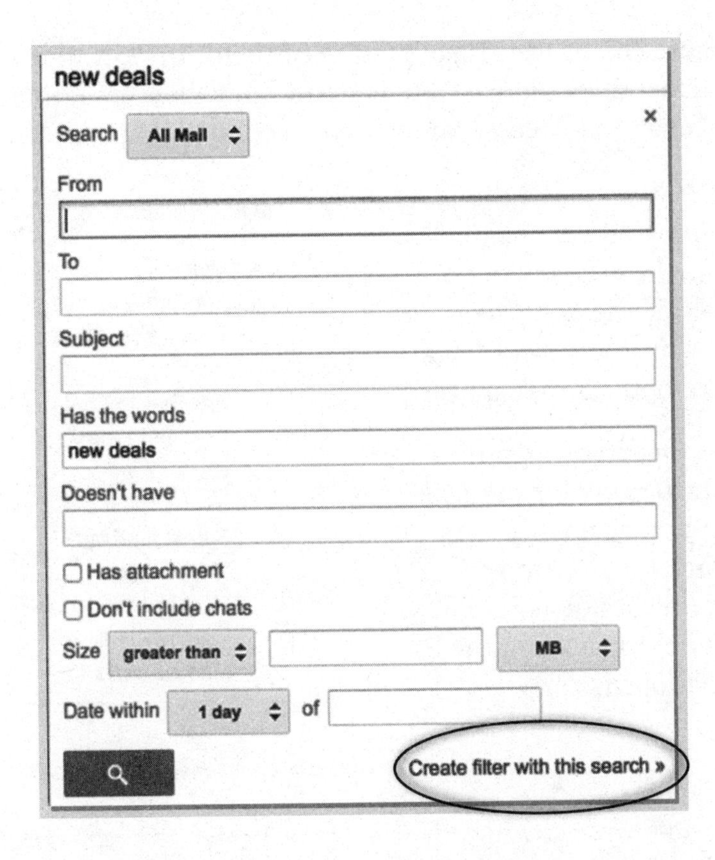

3. Click *Create filter with this search* in the bottom-right of the box to finish setting up your filter. This will take you to another window with 10 more options. Select those you want to apply, then click the *Create filter* button in the bottom-left corner.

Protect your Inbox from unwanted emails

Read this before you click that "Unsubscribe" button! Don't unsubscribe from an unwanted newsletter if it came from a company you don't recognize or don't trust. Sometimes opting out from these emails just triggers more of them. Luckily, there's a way to label them "junk" so they don't blow up your account. Here's what you need to know to protect your Inbox.

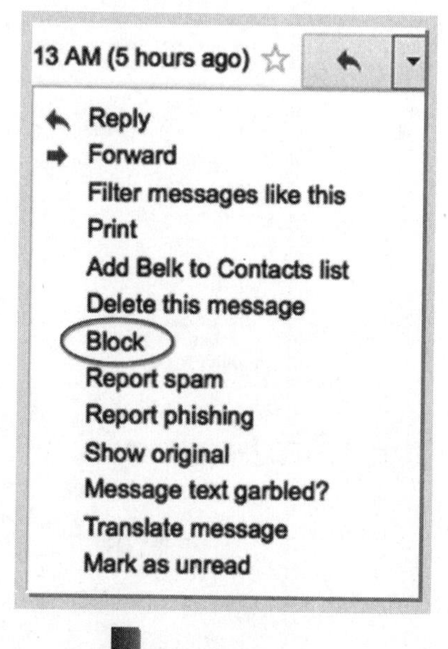

- In Gmail, open the annoying email and click the drop-down arrow beside the *Reply* icon. Then select *Block* from the options that appear.
- In Outlook, open the email, look for the *Junk* option at the top, and click the drop-down arrow beside it. Then choose *Junk* from the list that opens.

Easy way to dump the junk

So much junk mail, so little time. Isn't that the way it goes with email these days? Thankfully, you no longer have to scroll to the bottom of an unwanted email, searching for that itty bitty Unsubscribe link, to stop getting junk.

- In Gmail, open the email and look for the word *Unsubscribe* next to the sender's address at the top of the message.

- In Outlook, open the email, then click the arrow beside the word *Junk*. This will open a drop down menu with a few options. Click *Unsubscribe* at the bottom.

Lightning-fast way to clean up your Inbox

Here's the fastest way to clear junk emails from your Inbox. Sign up for Unroll.Me. It takes less than a minute and is so easy to use, it seems too good to be true.

When you register for a free account at *www.unroll.me*, you give the app permission to access your email account. It then looks for any subscriptions you receive and opens a window with three options — *Add to Rollup, Unsubscribe*, or *Keep in Inbox.*

- If you choose *Add to Rollup*, the app will place emails from that sender into a single email called the Daily Rollup. Any subscription you *Add to Rollup* will appear in this digest. You can even choose whether you receive your digest in the morning, afternoon, or evening.

- Choose *Unsubscribe*, and the app will take you off that sender's mailing list. You won't have to deal with the usual maze of questions that bombard you when you try to unsubscribe.

You can change these preferences at any time by changing your Unroll.Me settings. The service works with many popular email providers, including Outlook.com, Gmail, Google Apps, Yahoo! Mail, AOL Mail, and iCloud.

3 secrets to stopping spam for good

Just say "no" to spam — email spam, that is. Check *www.snopes.com* to see who is really behind all those spam emails. Once you know, you'll know how to stop them. (To learn more about Snopes.com, see page 138.) Plus, you can try these top ways to avoid getting it.

- Create one email account for personal messages and a different account for online shopping and any other websites that ask for your email address. That way, most of your spam will end up in your secondary account.

- Don't open a message from a suspicious or unfamiliar address. Because if you do, and you click any links in it, your email address could end up being sold to other spammers.

- Create a temporary email address at *10minutemail.com*. Your disposable address will last for only 10 minutes, long enough for you to send an email and receive a confirmation. Need more time? Click the *Give me 10 more minutes* link. Any replies you receive will show up on the website but disappear completely after your time is up. So take a screen shot of the reply if you need it for future reference.

3 ways to catch 'phish' before they hook you

Fall for a phishing scam, and you'll find yourself in a fine kettle of fish. That's because phishing scams are designed to look like legitimate emails, but they're not.

These deceitful messages look like they're coming from your bank, other financial institutions, or online services like eBay or PayPal. They tell you there's a problem with your account and ask you to click a link. Don't do it! Don't become a victim of this email trap.

A financial company will never send you an email asking for personal or financial information. Email is not a secure enough way to share this sensitive info. Here are three ways you can steer clear of these con artists.

- Never click a link in an email and then enter or update personal information like a password or credit card number.
- Do visit your account's website through your Web browser and log in normally. If there's truly a problem, you'll see it on the website.
- Don't fill out financial forms contained in email messages, and avoid using email to share a credit card or Social Security number.

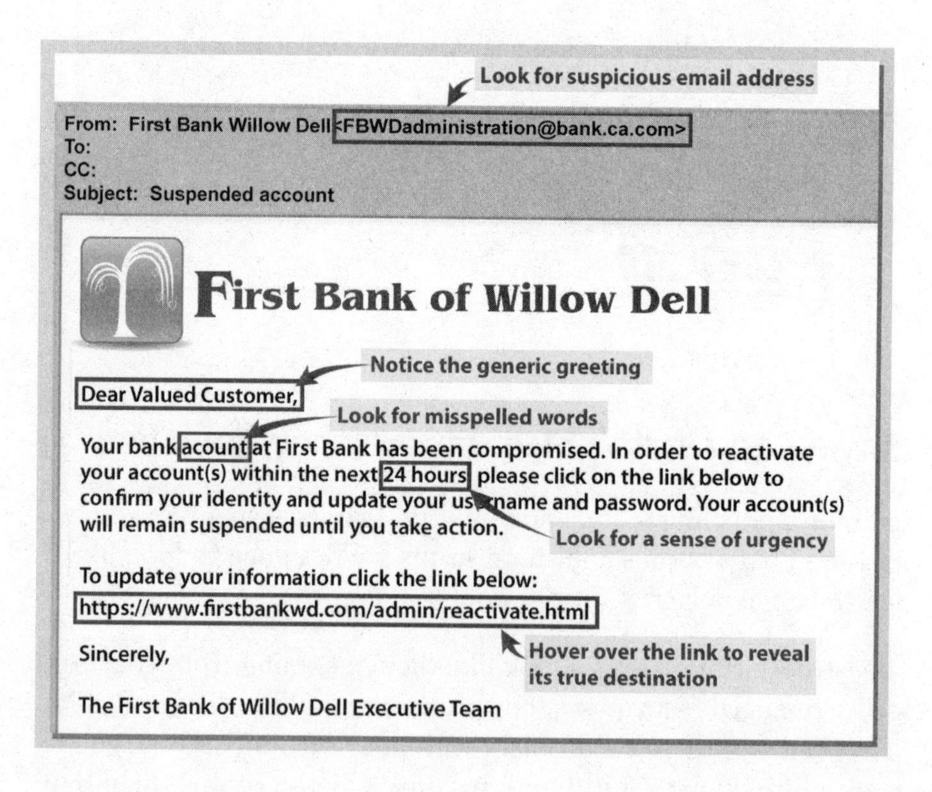

Keep your computer bug-free

To open or not to open? That is the question. Well, maybe not during Shakespeare's time, but certainly in today's Internet age. When you get an email attachment that says, "You're never going to believe this!" it's hard not to open it. But if you do, you could be opening the door to a computer virus.

One way to tell if an attachment is germ-free is to look at the file extension. The file extension is the three or four letter code that follows the period at the end of a file name, such as .TXT or .JPG. Attached files that end with .TXT and .JPG are generally safe, as are .BMP, .GIF, .PNG, and .TIF files. It's probably safe to open these if they came from a trustworthy source.

Other files are not so safe. Steer clear if an attachment ends with any of these extensions. It might harbor a virus.

.BAT	.HTA	.SCR
.CMD	.JSE	.PIF
.COM	.WSF	.VBE
.EXE	.LNK	.VBS

Bad guys may try to disguise a dangerous file by giving it two extensions — one safe and one malicious, such as ".JPG.EXE." Don't be fooled. The extension furthest to the right is the one your computer will try to open. And the one that will corrupt it.

This list may help you avoid catching a bug, but if you get an attachment from any suspicious source, don't take chances. Don't open the attachment. Simply delete the email from your Inbox and from your trash.

Snoop on Snopes to learn the truth about that suspicious email

Did you get the email about the Nigerian ambassador who wants to send you a check for $100,000? So did everyone else. How do you know if a tale like this one is truth or fiction? Check *www.snopes.com*. There, you can find out what's true and what's not in those crazy emails and stories making the rounds.

Snopes researches everything from urban legends circulated via email to photo-shopped pictures on social media. The husband-and-wife founders do their best to investigate thoroughly then report their findings. It's easy to use, too. Just type a topic in the search field and click *Go*.

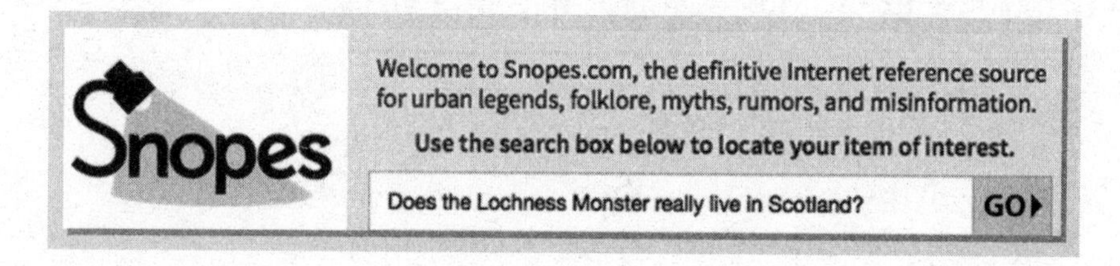

Before you pass along a far-fetched story, check with Snopes. And if you do decide to share that email, don't forward it. You may expose your friend's email address to spammers. Instead, copy and paste the text into a new email.

Email big pics without them bouncing back

Sending photos has never been easier, even if you think they're too big to email. Services like Outlook, Gmail, and Apple's Mail app have found ways to make them a snap to send.

- Apple's Mail app can make your pictures small enough to send by compressing them and placing them in a zip file, or by sending your recipient a link that lets him download the photos within a certain amount of time. Mail also lets you email your photos via Mail Drop to your iCloud account.
- Both Outlook and Gmail let you drag and drop plenty of photo files into your email without them bouncing back. But if you do happen to reach your size limit, both email providers let you share your photos with OneDrive or Google Drive.

Save yourself from sending an embarrassing email

This one little tip could save you endless email embarrassment. Don't fill in the recipient's email address until you finish writing the message. That way, you can write and revise to your heart's content. If you accidentally click send before it's ready, your email won't go anywhere. Only fill in the recipient's email address when you are confident that your message is fit to send.

Express yourself with personalized signatures

Here's a way to save a little time with every single email you send. Create a customized signature, then set up your email to automatically add it at the end of each message. You can sign your name, quote a great writer, or even insert a picture — whatever impression you want to make.

- In Gmail, click the gear-shaped icon in the upper-right corner of the window. Click *Settings*, scroll down to *Signature*, and fill in the field. Pick from a variety of fonts, sizes, and colors. You can even insert an image. Make sure you scroll to the bottom of the window and click *Save Changes* when you're done.

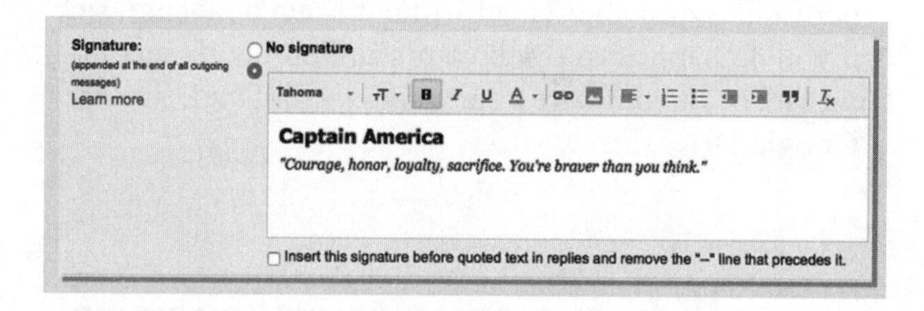

- In Outlook, click the gear icon in the top-right corner of the window. Click Options > Formatting, font and signature. Type whatever you like in the *Personal signature* field and jazz it up with a fancy font or color. Just remember to click the *Save* button when finished.

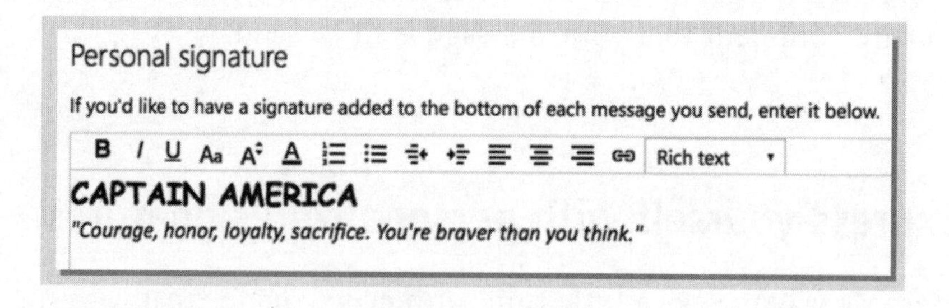

You can do this with other email programs, too. Look for signature choices under headings such as *Settings, Preferences,* or *Options.*

Tell your contacts you're on vacation — automatically

You're packed and ready to head out on vacation. There's just one last bit of business to take care of — alerting your email contacts. You can set up your email to respond to people when you're away. In both Gmail and Outlook, start by clicking the gear icon in the upper-right corner of the window. Then:

- in Gmail, click *Settings* and scroll down to *Vacation responder*. Fill in the starting date and type the message you want your contacts to receive in the *Subject* and *Message* fields. You can also enter an end date (*Last day*), or check a box so that only your contacts get your vacation message. Click *Save changes* at the bottom. If you don't fill in an end date now, remember to open *Settings* again when you return and click the circle next to *Vacation responder off*.

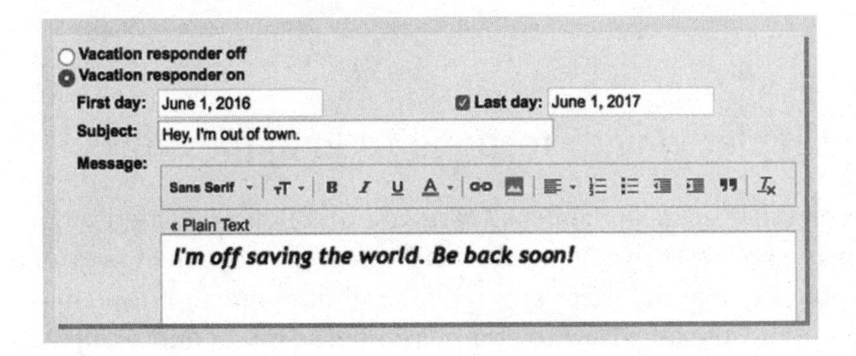

- in Outlook, click Options > Sending automated vacation replies, and then type the message you want people to see. Only people in your Contacts list will see this if they try to email you. You can have it sent to everyone who emails you by unchecking the box beside *Only reply to your contacts*, but understand you may receive more junk mail if you do this.

Click *Save* at the bottom of the page when finished. Outlook doesn't allow you to set an end date, so remember to go back in and click the circle next to *Don't send any vacation replies* upon your return.

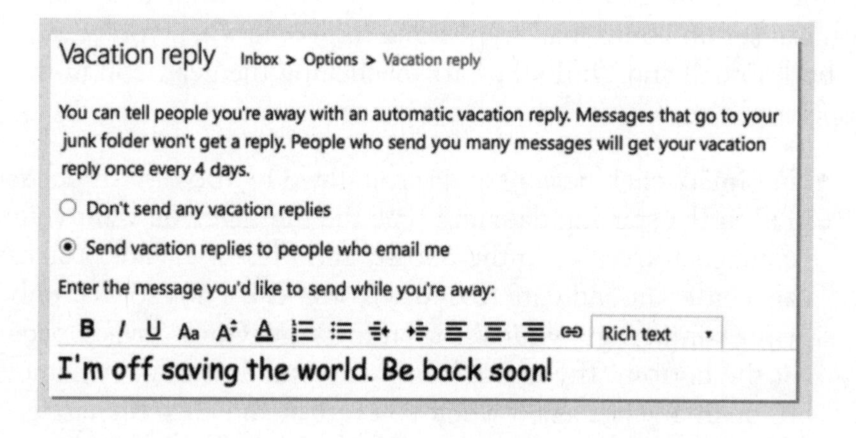

A simpler way to respond to long messages

Something magical happens when you highlight a portion of an email you've received, and then click *Reply*. Many email providers will paste only the part you highlighted into your response, making it a whole lot easier to comment on a small section of an otherwise lengthy message.

Not all email services do this by default, but you can change your email program's preferences so it does.

Family Reunion

It's that wonderful time of year again! I hope you're as excited as I am for our annual family reunion. All of the Avengers and their families should be present. Well, except for Loki. He's not invited. I wondered if you could take charge of any of the following duties for our little gathering.

1. Food and refreshments. We'll need plenty. You know how much Thor loves to eat. He's like a kid in a candy store. Only bigger.
2. Games. I was thinking sack race, egg toss, that sort of thing, but wasn't sure if Hulk could handle them. Thoughts?
3. T-shirts. We need a design. Maybe Iron Man can come up with something?
4. Family portrait. Do you know a good photographer with a wide-angle lens? It's the only way to fit Hulk in the photo.
5. Evening event. Could we all go see the new "Spiderman" movie? Those are always good for a few laughs. (As if a kid could ever shoot webbing from his wrists. Haha!)

I look forward to hearing from you.

Captain America

5. Evening event. Could we all go see the new "Spiderman" movie? Those are always good for a few laughs.

I'd love to see the new "Spiderman" movie. He's like a long-lost relative.

Black Widow

- In Apple's Mail program, click the word *Mail* at the top-left side of your screen and choose *Preferences*. Click *Composing* along the top of the window that opens, then click the circle beside *Include selected text, if any; otherwise include all text* at the bottom.

- In Gmail, click the gear-shaped icon in the top-right corner of the window, select *Settings*, then click *Labs* along the top.

General Labels Inbox Accounts and Import Filters and Blocked Addresses
Forwarding and POP/IMAP Chat Labs Offline Themes

Scroll down to *Quote selected text* and click the circle beside *Enable.*

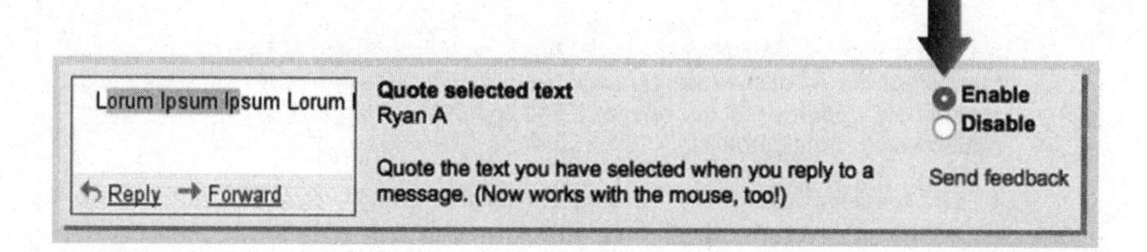

Make sure you save your preferences by scrolling to the bottom of the page and clicking the *Save Changes* button. If you have a different email service, rummage through your settings, options, or preferences to see if you can set up this feature.

Send group emails while protecting people's privacy

It's probably one of the most annoying things to come through your email — group messages with dozens of people's email addresses stacked a mile high, like a triple-decker sandwich.

All those addresses out in the open also pose a privacy risk. This is where the bcc feature comes in handy.

Bcc stands for blind carbon copy, and it's a way to send a lot of people an email without the recipients seeing one another's addresses. The trick is to type your own email address into the *To* field, and type everyone else's into the *Bcc* field. The result — only you will see all of the recipients; the others are "blind" to one another.

Here's how to make the *Bcc* field appear when you write an email.

In Apple's Mail program, click *View* in the toolbar along the top of the window. Then select *Bcc Address Field* from the drop-down menu. From now on, the *Bcc* field will appear below the *To* field in each email.

In Outlook, click *+ New* to start a message, then click *Bcc* above the *To* field.

In Gmail, click the *Compose* button to begin a new message, then click *Bcc* at the end of the *To* field.

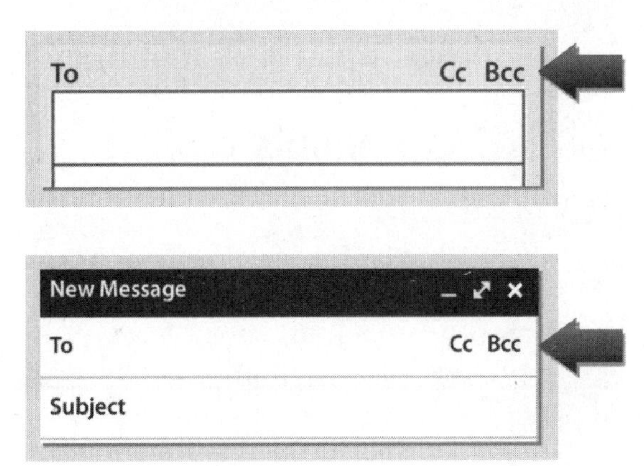

Save the date with a single click

It's time to retire your bulky day planner. The next time you get an invitation via email, hover over the date with your mouse pointer. The email service will add the event to whichever calendar it's linked to. For instance, Apple's Mail program adds events to your

Apple Calendar; the Windows 10 Mail app to your Windows Calendar; and Gmail to Google Calendar. Either way, fill in the details, and you'll never forget a lunch date again.

Here's how it looks in Apple's Mail app. Your email service may look slightly different. Make sure you click *Add to Calendar* (Apple Mail app, Gmail) or *Save and close* (Windows 10 Mail app) when finished.

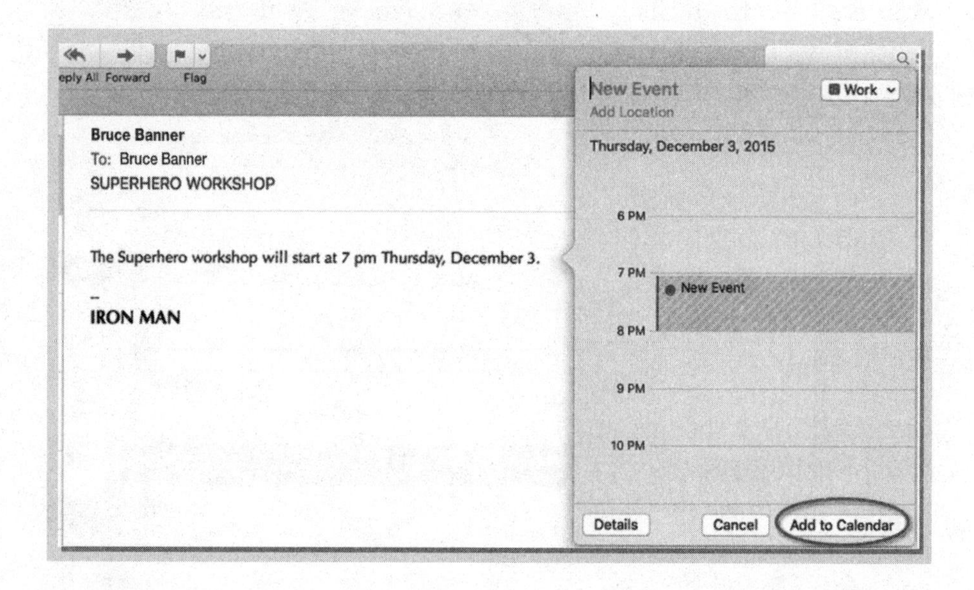

Save face by 'unsending' an awkward email

Oops! You hit send and didn't mean to. It happens to the best of us, but if you use Gmail, you're in luck. Google's email service now offers an "undo" feature that lets you cancel an email within 30 seconds of sending it.

Click the gear-shaped icon in the upper-right corner of your Gmail window, then click *Settings*. Under the *General* tab, scroll to *Undo Send* and click the box beside *Enable Undo Send*. Pick the delay you want, up to 30 seconds, by clicking the drop-down list after *Send cancellation period*. Remember to scroll to the bottom of the page and click *Save Changes*.

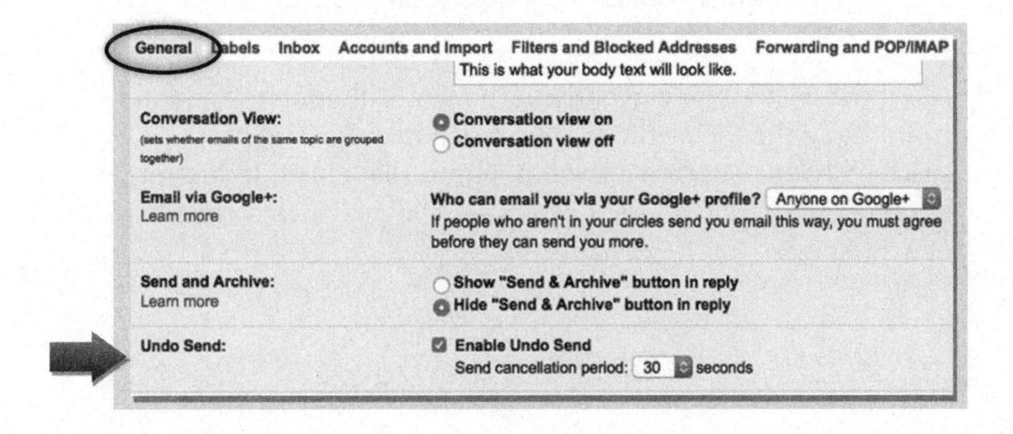

Now every time you send an email, Gmail will automatically show you this message. Simply click to undo or review your message.

Your message has been sent. Undo View message

Easy way to attach files on a Mac

Here's a super-secret way to add an attachment to an email on a Mac computer. Start composing a new email, then open Finder to find the file you want to attach. Drag it to the Mail icon on your Dock.

All of the emails you currently have open will jump to the forefront. Drag the file to the correct email message and press your space bar. Your attachment will pop into that email. If you forget to press the space bar, don't worry — Mail will attach the file to a new message.

Smartphones & tablets

Make the most of your gadgets

Fit more apps on your phone's screen

Your phone's Home screen doesn't hold many icons. Instead, you have to flip through page after page of apps almost every time you open your phone. There's got to be a better way, right? On some Android phones, including the Samsung Galaxy S6, there is.

See your most important apps right away. Sometimes the apps on your Home screen aren't the ones you use the most. Changing that is simple. First, tap *Apps*. Then press and hold the app you want, and drag it to a spot on your Home screen. You can remove rarely used apps from your Home screen the same way. Just press and drag them to another screen.

Make room for more apps. This trick shrinks each icon in order to fit more of them on the screen.

1. Press and hold a blank area on your screen.
2. Tap *Screen Grid* among the options that appear at the bottom.
3. Decide how many icons you want to see at a glance, and tap that option at the bottom of your screen. For instance, tap *4 x 4* if you want 16 icons per page; *4 x 5* if you want 20; or *5 x 5* if you want to see 25 icons at a time. That's 50 percent more icons than on a 4 x 4 screen!
4. Tap *Apply*.

Rely on widgets for fast info. Widgets can save you time and hassle by showing important information like the weather, date, emails, and even stock prices on your Home screen. You can see them at a glance without having to open an app.

Your phone comes stocked with widgets that you can add with just a few taps. On a Galaxy S6, press and hold a blank space on your screen, and tap *Widgets* at the bottom. Swipe through the available widgets until you find the one you want. Press and hold it, then drag it to an empty space on your Home screen and drop it.

Get longer battery life than ever before

You can almost see your battery indicator dropping right before your eyes, so you're beginning to wonder if you have just one energy hog — or a whole herd. Now you can finally get some answers.

On Apple devices, tap the gear-shaped Settings icon. From the Settings screen, scroll down and tap *Battery*.

Apple Settings

On the battery screen, scroll to see what percentage of battery life each app has used in the last 24 hours. Don't be surprised if your Home & Lock Screen is one of the biggest offenders because screens can be surprisingly power-hungry.

You may not have control over your energy-gobbling apps, but you can save power on your Home screen by making it dimmer. Lowering the brightness level can make your battery last a lot longer.

Try taking it down to only 50-percent brightness, and give yourself some time to adjust to it. You may eventually become comfortable with the darker screen. Just make sure you can see it well enough to read without struggling.

Don't use the automatic brightness setting — that actually uses more power than manually dimming and brightening your screen. To dim the screen yourself, follow these steps.

On your iPhone and iPad. Tap Settings > Display & Brightness. If the *Brightness* slider is near the right, drag it left to dim your screen. If the *Auto-Brightness* slider is green and turned on, drag the slider left to turn it off.

On your Android phone and tablet. Swipe down from the top of the screen to lower your Notification panel. The sun with a line next to it is the brightness indicator. Drag the brightness slider left to dim the screen to a comfortable level. If the *Auto* checkbox is filled, tap it to turn it off.

You can also access the brightness indicator on your phone by tapping the gear-shaped Settings icon, then *Display*. On your tablet, tap Settings > Device > Display > Brightness.

Android Settings

Limit screen updates to save battery power

If your battery life seems to get worse every time you install a new app, push notifications may be the reason.

You may love these little updates and messages that appear on your screen, but your battery doesn't. Too many push notifications can suck away battery power like a vacuum cleaner on steroids.

The problem is, an app can send them even when it's closed and you're not using your phone. These notifications "wake" the phone, trigger the notification chime, and flash a message on your home or lock screen to get your attention.

So if you already struggle with keeping your phone charged, be careful when you add a new app to your phone. While installing and setting up the app, watch for any opportunity to opt out of push notifications. It's easier to turn down push notifications now than to get rid of them later — and your battery will thank you.

Pump up your battery with the flick of a switch

Darn. You need to check messages, but your battery is almost dead. To eek out the last bit of power before you can recharge, turn on your device's power-saving mode.

On your Android phone. Tap Settings > Battery. If you don't see any power-saving modes on the battery screen, tap Settings > Power saving instead.

- To save a little power, tap *Power saving mode*, and turn on the slider. On the Galaxy S6, this triggers power-saving tricks that make the battery last up to 10 percent longer, without putting heavy restriction on your phone's performance or the look of your screen. When you want to turn it off, swipe down from the top of your screen to view your Notification panel, and tap *Power saving mode on.* Drag the *On* slider left to turn it off.

- If you're desperate to stretch the phone's battery life, tap *Ultra power saving mode.* The screen changes to a black-and-white theme, just like an old-fashioned television. The device also limits its performance and how many apps you can use. To return to normal power settings, swipe down on the Notification shade, and tap *Turn off Ultra Power Saving Mode.*

On your Android tablet. Tap Settings > General > Power saving.

- Locate *Power saving mode* and turn on its slider. It's on when the toggle turns green.
- To select Ultra power saving mode, drag its slider right. When the *Ultra power saving mode* dialog box appears, tap *Turn on.* When you want to escape from Ultra power saving mode, swipe down from the top of the screen to make Notifications appear, and tap the *Turn off Ultra power saving mode* notification.

On your iPhone. Tap Settings > Battery. Turn on the *Low Power Mode* slider. Apple says this should give you another three hours of battery life. Plenty of time to check those messages!

Charge your device faster than ever before

You take out your tablet to dig into your favorite novel and the battery is about to die. What gives? You just charged it. While charging your tablet or phone is as simple as plugging it in, you can get that battery icon to 100 percent much faster by following a few simple guidelines.

To start, turn off your device. If it's on and you use it while charging, more juice may be going out than coming in. If your device stays on while charging, you can reduce the amount of power being used by putting it in Sleep mode or Airplane mode.

Sleep mode puts your tablet on standby, while Airplane mode keeps a phone or tablet from sucking up power to connect to Wi-Fi, cell service, or Bluetooth.

To get the fastest charge, use the cable that came with your tablet or phone, and plug it into the wall using the AC adapter. While you can charge a device using a USB port on your computer, it will go much slower.

New rule — charge your battery often

Do you wait until your phone is nearly dead before charging it up? Uh oh. You may be dooming your battery to an early grave.

Once upon a time, your phone's battery life depended on following that rule of thumb. As recently as 2007, phones were stocked with nickel-based batteries that could only be charged a certain number of times before they wore out.

But today's phones have lithium batteries. They fare better if you charge them before they dip below 50 percent. So you need to do just the opposite of what you're used to.

Also, try not to keep your phone plugged in overnight. Leaving it connected to a charger for long periods can shorten battery life, too.

Mute your phone fast with a single touch

You're attending that once-in-a-lifetime mass led by the Pope when your phone shatters the silence. Embarrassing moments like that are not the time to be frantically trying to turn off your phone. Fortunately, you can quickly mute the ringing without disconnecting the caller.

On your iPhone. Press the Volume Down button on the side of the phone once, and silence will fall like snowdrifts. Your caller won't know the difference — he'll still hear the ringing until the voicemail message starts.

Keep in mind this may not work if you forward your calls to other Apple devices. If they're nearby and on Wi-Fi, they may also pick up the call.

On your Android phone. Press one of the Volume buttons on the side of the phone. Easy as pie.

Stop your screen from flipping around

You can turn your phone or tablet sideways to watch a video, but tilt your device just a little too far and the screen turns "right side up" again — cutting off part of your video.

If you're using your tablet, you can lock your screen to stay either vertical (portrait) or horizontal (landscape). Your phone will only lock in portrait mode.

Android
Rotate

On Android phones and tablets. Swipe down from the top of the screen to access Quick Settings. Tap the Screen rotation icon to lock it.

When you're ready for your phone or tablet to return to its spinning ways, just tap the icon again to unlock it. The icon will turn yellow to show screen rotation is in use.

On iPhones and iPads. Swipe up from the bottom of the screen to open the Control Center. On the top-right, tap the padlock icon to lock the screen. The icon will turn white when screen rotation is locked. Tap the icon again to turn off the screen-locking option and let your screen rotate freely.

iOS Rotate

Host a successful conference call anywhere

Can you talk to more than one person at the same time on your cellphone? Yes, if your phone carrier allows it. Just check first to be sure conference calls are possible on your network and that the calls don't cost extra.

Here's how you host your first teleconference, whether it's on an Android phone or an iPhone. Say you want to call Sonny in Phoenix and Wendy in Chicago.

1. Dial Sonny's number.
2. After he picks up, tap the plus-shaped *Add call* icon.
3. Dial Wendy's number, either with the keypad or by tapping her name in *Favorites* or *Contacts.*
4. When you call Wendy, Sonny will be put on hold. To bring him back into the conversation and start the teleconference, tap the *Merge* or *Merge calls* icon. Now you can all hear and talk to each other as if you were in the same room.

5. To add more people to the teleconference, repeat steps 3 and 4 for each one. But be aware your carrier or phone manufacturer may limit the number of people you can include in a conference call.

To end the teleconference, press the phone-shaped End Call or Receiver icon to hang up on everyone at the same time. Hopefully, you told them goodbye first.

Timesaving tip: set friends and family as favorites

Forget searching for contacts you call often on your iPhone. Find them fast with these tips.

Pull up your most recent calls. Tap your Phone icon. Tap *Recents* at the bottom of the screen to see all the calls that have recently come in. Tap any name or number in that list to call the person.

Pick your favorites. Get quick access to friends and family by putting them in your Favorites list. Here's how.

1. Tap the Phone app icon just as you would when making a call.
2. Tap the star-shaped *Favorites* icon at the bottom of the screen.
3. On the Favorites screen, tap the plus sign in the upper-right corner.
4. Tap the name of the contact you want as a favorite. If the contact has more than one number, tap the phone number you prefer.

When you want to call a Favorite, tap the phone icon, then *Favorites.* Tap the person's name, and the phone will immediately start dialing.

Forward important calls to another number

You're expecting an important call on your cellphone, but it needs charging, and you have a shopping trip planned with your sister. "Just forward your calls to my phone while we're out," your sister says. But how do you do it — and is it free?

Before forwarding any calls, check your phone carrier agreement and website to make sure they won't charge extra for this perk. Some plans may charge for this service while others don't, so check your plan specifically.

Also find out whether you forward calls with your phone's main Settings app, the settings inside your dialer or Phone app, or by using a call forwarding code such as *72.

Although the methods for forwarding a call vary, this is roughly how it should work.

- Navigate to your call forwarding settings or tap your carrier's call forwarding code.
- Type in the number to forward your calls to.

After that, the process depends on your carrier and phone. You may need to follow more instructions or choose which calls are forwarded. Your phone manual or carrier's website can help. They can also help you deactivate call forwarding when you no longer need it.

Redialing tips you need to know

Murphy's Law suggests you're much more likely to get a phone call when your hands are covered in soap or they're too full to grab the phone. So use this trick to make calling back easier.

To start, tap the Phone app. If you don't see the keypad screen, tap the *Keypad* icon. On the Galaxy S6, it's an unlabeled set of dots in the lower right corner. On the iPhone 6, the *Keypad* icon appears on the bottom of the screen.

If no number appears on the keypad screen, tap the call button (phone icon) to see the last number called. If you recognize the number, tap the button again to make your call.

If you don't recognize the number, do a Google search on the number before calling. Some scammers call and hang up immediately just to trick you into calling them back.

Block unwanted calls on your phone for free

Stop getting repeated calls from crank callers, telemarketers, or your soon-to-be-ex-husband. Many phones offer a number-blocking feature at no cost to you. Why put up with nuisance calls if you don't have to?

On your iPhone. To block an individual number on your iPhone 6, follow these steps.

1. Tap the Phone app icon.

2. At the bottom of the screen, tap *Recents* if you've gotten a call from that number recently, or *Contacts* if the number is in your list.

3. When you find the number in *Recents*, tap the nearby i at the right. In *Contacts*, tap the contact's name.

4. Scroll down and tap *Block this Caller*.

5. Tap *Block Contact* to confirm.

Finally, rub your hands together gleefully because you won't be hearing a ring from that number again.

On your Android phone. If you own a Galaxy S6 or similar phone, follow these steps.

1. Tap your Phone app icon.

2. Tap *Log* or *Contacts* at the top of the screen.

3. Tap the phone number you never want to hear from again.

4. Tap *More* in the upper right corner.

5. Tap *Add to Auto Reject List*. From now on, the call goes straight to voicemail without ringing.

Check your carrier's limitations. Some carriers may charge for certain call-blocking features or for blocking too many lines. Contact your carrier if you have any questions.

Discover when messages were sent or received

The message in your iPhone Messages app says "Meet me at the shopping center in one hour," but you have no idea when the message was sent. Have you already missed the meeting?

There's an easy way to find out. Press on the message and drag it to the left. You'll see time stamps for all your messages on the right side of your screen.

See, that message was sent just 10 minutes ago. You can still make it.

Forward that funny text message

Did that last text message make you laugh so hard that both cats fled the room? Here's how you can share the laughs with your friends, before your cats return for revenge.

Press and hold the text message you want to share. Then, if your phone is:

- an iPhone, the options bar appears. Tap *More*, then tap the ⮫ in the bottom-right corner.
- an Android phone like the Galaxy S6, tap *Forward* in the menu that appears.

Both phones copy the original text into a new message, which you can now send to other people. Just type the names or phone numbers of your friends in the *Recipient* or *To:* field, then tap *Send*.

Escape an annoying group text

Famous NFL running backs like Todd Gurley and Melvin Gordon love taking part in big, group texts with other NFL players. But group texts aren't so great when a parade of (mostly useless) messages makes your iPhone buzz or beep every five seconds. Take back control with these tips.

Create your own cone of silence. Maybe you don't mind getting the messages, but you're tired of the notifications and noises that come with every single text. Open iMessage, make sure you are in the conversation you'd like to mute, and tap *Details* in the upper right corner.

Turn on *Do Not Disturb*, and you will no longer hear sounds each time you receive a message in this conversation. This will last until you turn off *Do Not Disturb*. You'll still receive the messages, but they won't interrupt you. You can check them later at your convenience.

Leave the conversation. If you were chatting in person, you could slip away from the group. The iPhone lets you do the same thing. Keep in mind that once you exit, you cannot reenter the conversation. Plus, the other people in the group will know you chose to leave. Still want out?

1. Open iMessage, and make sure you're in the conversation you want to escape.
2. Tap *Details* in the upper-right corner.
3. Tap *Leave this Conversation* at the bottom of the screen.
4. Confirm that you want to exit the group by tapping *Leave this Conversation* in the dialog box.

When you do this, other participants will get a message saying you left the chat. You may also get a message confirming you left. What you won't get are any more messages from the group conversation.

It didn't work. Sometimes, *Leave this Conversation* is gray and won't respond, or you may not see it at all. This may happen because:

- someone in the group is sending texts with an app other than iMessage, or is using an iPhone that runs iOS 7 or below.
- only two other people are participating in the group conversation. You can't leave a group message that only has three people in it. It must have four or more.

In either case, you can't leave. Use the earlier instructions to turn on *Do Not Disturb*, instead.

Battery saving trick — use Wi-Fi in Airplane mode

Turn on Airplane mode, and your phone or tablet automatically turns off Wi-Fi, your cellular signal, and Bluetooth. You're required to do that while flying (that's why it's called Airplane mode!), but some frequent fliers use Wi-Fi while Airplane mode is on. And guess what? You can use a similar trick to stretch your phone's battery life.

Even when you're not using your phone, it's working hard searching for cellphone towers and scanning for Wi-Fi networks. In fact, your phone never stops looking for a cellular signal, even when you're in places with little or no coverage. All that searching uses a lot of battery power.

If you're OK with not getting phone calls or texts for a while, you can turn on Airplane mode to save your battery, then restart Wi-Fi to work online. And, of course, you can also do this while flying certain airlines if you're willing to pay the Wi-Fi cost. Here's how to set it up.

On your Android phone or tablet. Swipe down from the top of your screen to open your Notification panel, then follow these steps.

1. Tap the grid (tablet) or pencil icon (phone) in the upper-right corner to open your full Quick Settings menu.
2. Tap the airplane icon to turn on Airplane mode, and follow the prompts.
3. Turning on Airplane mode turned off Wi-Fi, so tap the grid or pencil icon again.
4. Tap the Wi-Fi icon to turn wireless back on.

WiFi

When Airplane mode is on, an airplane icon appears near the
top-right corner of your screen. That will help remind you to turn it off when you need to receive calls and messages again.

On your iPhone or iPad. Swipe up from the bottom of the screen to open the Control Center. Tap the airplane icon to turn Airplane mode on. After it turns white, tap the Wireless icon to turn it back on. The airplane icon displays in the upper-left corner of your screen and remains until you turn Airplane mode off.

Should you make your phone a Wi-Fi hotspot?

You've heard hackers can break into free Wi-Fi in public places, so you wonder about using your phone as a more secure Wi-Fi hotspot for your laptop. Before you try this, here are four things you absolutely must know.

What exactly is a hotspot? Your smartphone has its own Internet connection, so you can connect your laptop to the Internet through your phone via Wi-Fi, Bluetooth, or a USB cable. It's called tethering.

Most smartphones can do this if your service provider allows it. If you set up the connection through Wi-Fi, your phone becomes a Wi-Fi hotspot. That means you just log in like you do with any wireless network.

Experts say it isn't hard to do and can be very convenient. But it comes with one big fat warning.

Beware of skyrocketing charges. Because tethering can devour a lot of data, it quickly gets expensive if you go over your cell plan's data limits. Occasionally checking email and browsing the Web may use about two or three gigabytes (GB) of data. But long periods of tethering or short periods of watching video will most likely blast through your data plan limits.

Before you try tethering, see the next tip to learn how to monitor your data use and help prevent unpleasant phone bills. And be aware that excess data charges may not be the only surprise.

Check your carrier's rules. Some carriers could shock you with a high bill if you use your hotspot without understanding their rules, while others may prevent you from turning on your hotspot settings without permission.

So, before you tether, ask your carrier how to turn your smartphone into a hotspot and how much it costs. You might even check their terms and conditions online to make sure you know what to expect.

Each carrier handles tethering costs differently. While some only make you pay for data use above your plan's limits, others may require you to switch to a new phone plan or buy a hotspot data plan on top of your regular data plan. Just understanding how things work may save money.

Give your battery a boost. Tethering can also cost battery life because the phone uses up more battery power when serving as a hotspot — sometimes a lot more. Try plugging your phone into a charger, laptop, or other power source to help.

Data tracking — easy way to avoid overage fees

Remember the shock of that shower scene in *Psycho*, where Janet Leigh sees the knife outside her shower curtain? Go over your phone plan's data limits, and opening your bill may make you feel the same shock and horror.

Data overages are costly enough to make anyone scream. But spend five minutes making the right adjustments to your phone, and you may never go over your data limits again.

On your Android phone. You'll find everything you need to know about how much data you're using under Apps > Settings > Data usage.

- To see the current amount, look at the date range. The dates should match your carrier's billing cycle, but if they don't, tap the dates, and tap *Change cycle* to set the correct dates. If the date range is correct, check the number next to it to see how many megabytes or gigabytes of data you've used since the billing cycle started.

- To set a limit on your mobile data, tap *Limit mobile data usage* to turn its slider on. On the graph below, find the red slider adjustment bar, and drag it to set a limit. It's a smart move to choose a number below your data plan limit. Now your phone will switch off mobile data when you reach the number you set.

- Need advance warning you're about to go over? Tap *Alert me about data usage* to turn its slider on. As before, a slider adjustment bar appears on the graph. Drag it up or down to set an amount below your data plan limit.

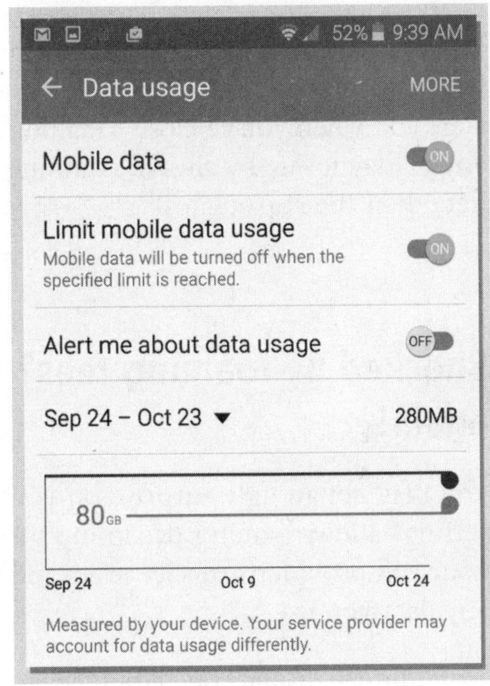

On your iPhone. To find out how much data you've been using, tap Settings > Cellular. Under *Cellular Data Usage*, you'll see an amount for *Current Period.*

Don't panic if it's a really large number. It's probably because you've never reset the counter, and you're viewing all the data you've used since you bought your phone. Not helpful.

To stay on top of your data usage, you need to reset the counter at the beginning or end of each billing period. Simply scroll down to the bottom of the screen and tap *Reset Statistics.* Now set yourself a reminder to do it every month. If you tend to be forgetful, an app may help.

Your phone carrier may offer a data-usage app, but you need to monitor it closely to make sure the numbers update often enough to stay current.

You can also try a free app called My Data Manager. Supply the app with information about your billing cycle and data cap, and it will alert you when you're close to falling into the dreaded data-overage abyss. Play it safe by always assuming you have used a little more data than the app suggests.

Shocked by roaming fees? Never be ambushed again!

You may get an ugly surprise on your cellphone bill if your phone settings allow roaming. Roaming means you've wandered outside your cell provider's service area, and you'll pay a huge price for any data you use.

Some international travelers have been shocked to receive bills with roaming fees in the thousands. To help avoid a similar fate, learn how to keep track of this extra charge before it can shock your wallet, too.

On your iPhone. To turn off data roaming, tap Settings > Cellular > Cellular Data Options > Roaming, and slide the Voice & Data Roaming button to the left.

You should also check whether you've accumulated roaming data before now. Tap Settings > Cellular and look for *Current Period Roaming*. What you see there may cover a few days of roaming or a few years, depending on if you've ever reset your phone.

You learned how to reset the counter in the previous tip by tapping *Reset Statistics*. Once it's down to zero, recheck *Current Period Roaming* occasionally to see if any new roaming data has snuck in.

On your Android phone. To turn off roaming, tap Apps > Settings > Mobile networks > Data roaming > Roaming network. Tap the button for *Home only*.

To make sure all roaming settings are turned off, check occasionally to see if any roaming data has accumulated. Some carriers may include this information in the monthly bill, but a free app can help if you'd rather find out before charges pile up. To try this, download the My Data Manager app from the Google Play store.

On any phone. If you still have roaming data, even after turning off data roaming settings, try turning off *Background App Refresh* for your apps. (For more information on how to do this, see the next tip.)

You can also take the extra precaution of putting your phone in Airplane mode. This is especially helpful if you're traveling out of the country and want to be absolutely sure you'll have no surprises when you return.

Shut the tap on data guzzlers

Just because you haven't opened an app doesn't mean it's not gobbling up data. What? That's right, a closed app can still check for new information, messages, and app updates, and some spend a lot of time doing it.

Apps that handle email, social media, weather, and news are the most likely offenders. But you don't have to sit by helplessly while these data-guzzling apps drain your data plan. Here's how to limit them so they only use cellular data when they're open.

On your Android phone. Tap Apps > Settings > Data usage, and scroll down to the list of individual apps. The number beside each app shows how many megabytes or gigabytes the app has used during the current billing cycle. Tap any of the apps you wish to restrict. To make sure the app only downloads new data while open or on Wi-Fi, turn on the slider for *Restrict background data.*

Some apps may allow you to do even more. Tap *View App Settings* to check for other ways to restrict how much data you use.

On your iPhone. To see which apps devour the most data, tap Settings > Cellular. Scroll down to *Use Cellular Data For* to see all your apps listed in alphabetical order. Beneath each app's name is the number of kilobytes or megabytes it has used since you last reset your data.

If you don't want an app to use cellular data, even when you have the app open, turn the slider off. The app can still send and receive data when you are signed in to a Wi-Fi network.

If you prefer to limit how much cellular data an app uses, tap Settings > General > Background App Refresh. Turn on the

Background App Refresh slider at the top of the page to activate all the apps below. Then turn off the specific apps you don't want operating behind the scenes.

4 ways a cellphone can save your life

Nearly 1,500 people were missing after the 2011 tornado that devastated Joplin, Missouri — and most communications were down. But one man trapped under the rubble of a building could still send his best friend a text message. That message helped rescuers find and save him.

The FCC recommends calling when you need to reach 911 and other emergency services. But if you don't have enough charge or if calls can't get through, try texting a friend or family member like the man in Joplin. Texting is also best for non-emergencies during a disaster because cellphone networks may be too crowded for calls to get through.

What other safety features are hiding on your phone? Finding out could save your life!

Make ICE your safety net. You've fallen off your bike and hit your head. When emergency workers arrive, will they know who to contact if you're unresponsive? They will if you've created an ICE (In Case of Emergency) entry in your phone.

If your iPhone runs IOS 8 or later, you're lucky — you have a Health app where you can set up emergency contact information. (See next tip for how-to's.) On a Galaxy S6, you'll find an ICE category in your Contacts list.

You can also download an emergency-contact app from iTunes or Google Play. Look for an app with lock-screen access if you use a passcode to lock your phone.

Of course, you can always do it the old-fashioned way — just write your emergency numbers on a sticker on the back of your phone.

Take advantage of your phone finder. iPhone's Find My Phone app and Android's Device Manager app may help save your life if you've set them up. (Learn how to do this on page 198.) Make sure your emergency contact or a close friend or family member also knows how to use these apps to find you.

Map a path out of danger. Learn how to use the GPS or map apps on your phone now, and they may save your life later. If you ever get lost in a dangerous location, these apps can help you find your way to safety.

Always have your phone ready. Nobody expects a sudden emergency, but they happen to someone every day. Keep your phone within reach, so you can quickly call for help if the worst happens.

Remember, if you swipe your iPhone's lock screen and tap *Emergency*, you'll have instant access to a keypad to dial 911. On an Android, swipe the phone icon in the corner.

Your phone could be a lifesaver — even if it's locked

"Hey Siri, tell the paramedics what they need to know to help me." Well, maybe Siri can't do that, but your iPhone Medical ID can. Emergency responders can use this app to quickly get vital information needed to help you during a medical emergency — even if you can't answer their questions.

Before emergency responders can use this app, you must set it up. Tap the Health app icon to get started. (It's a white square with a red heart.) To make sure medical professionals can see your information without your passcode, turn on the *Show When Locked* slider. Then tap Medical ID > Create Medical ID. Now you can type in crucial details like:

- your birth date.
- medical conditions.
- allergies.
- medications.
- emergency contacts.
- weight and height.

Use the *Medical Notes* field to list information like your preferred hospital or doctor, surgeries, implanted medical devices such as a pacemaker, and other health details.

Tap *Done* and close the Health app to save the information you've added. If you need to go back in to make changes, simply tap *Edit*.

Now, how does someone access all this information without getting into your locked phone? Here's the exciting part. Simply swipe right to bring up your passcode or lock screen. Instead of typing in a code, tap Emergency > Medical ID. It's that fast and easy. And, according to medical experts, the faster first responders can get important information about you, the better off you'll be.

The bad thing is that your medical information is now available to anyone who grabs your phone — including identity thieves and scammers. If you're concerned about privacy, you can turn off the *Show When Locked* setting, but your emergency info will no longer show up.

Replace your (lost, stolen, damaged) smartphone for free

Insurance for your phone? Don't do it! At least not until you check the terms of your credit cards. Some cards offer smartphone replacement insurance if you use that credit card to pay your monthly cellphone bill. If you find a card that does, here are some questions to ask.

- Does the insurance cover all missing phones or just those that are stolen or damaged?
- What proof must you provide to get payment on your claim?
- Do you have to pay a deductible?
- How much coverage does your card provide?

If the insurance offered by one of your credit cards sounds worthwhile, start using that card to pay your cellphone bill. If your phone is ever lost, stolen, or badly damaged, that one change could provide a big payoff when you need it most.

Shortcut magic cuts typing time

Cut time and typos when you text using this easy shortcut. Instead of typing out long words or phrases, create your own abbreviations that instantly bring up the words you want. No more typing long email addresses or favorite phrases. Here's how to make this magic happen on your phone or tablet.

On your iPhone or iPad. To add a new text shortcut, follow this example using the name Winchester-on-the-Severn. This Maryland town boasts a tie for the longest place name in the United States (shared with Washington-on-the-Brazos, Texas).

If you live in either of these places, you definitely need this shortcut!

1. Tap Settings > General > Keyboard.
2. Tap *Text Replacement*, and tap the + sign in the upper-right corner.
3. On the line labeled *Phrase*, type in "Winchester-on-the-Severn." Type your abbreviation on the *Shortcut* line. For example, you might choose "wos."
4. Tap *Save*, and the new entry appears in your list of text shortcuts.

Now when you type the shortcut, your phrase automatically replaces it after you tap the space bar.

On your Android phone. Here's how to take advantage of this helpful feature on the Galaxy S6.

1. Tap Apps > Settings > Language and input > Samsung keyboard.

2. Under the *Customization* heading, tap *Text shortcuts*. Tap *Add* in the upper-right corner.
3. Under *Shortcut*, type the text shortcut or abbreviation you want, such as "wos." Under *Expanded phrase*, type in "Winchester-on-the-Severn."
4. Tap *Add*, and the new shortcut appears in your text shortcut list.

Now whenever you type the shortcut and tap space, the expanded phrase will appear. It will also show up in your predictive text, so you can tap that entry if you prefer.

Hate typing? Use your voice instead

If you're not the fastest texter in the world, you'll love this feature that will have you texting as fast as you speak.

Voice typing converts your speech to text in many apps. So now, you can simply speak and, like magic, the words appear on the screen. It's especially handy when you're driving, as long as you set it up beforehand.

No matter which device you use, remember these keys to voice-typing success.

- Before you start speaking, open the app you want to type in, and tap where you want text to appear.
- Look for the microphone icon on both Android and Apple devices. That's your ticket to voice typing.

- Enunciate clearly to avoid errors.

- When you need to end a sentence, say "period" to type a period. You can also say question mark, comma, exclamation point, or colon to make that punctuation appear on the screen. To start a new paragraph, say "new line."

Now that you know the basics, try voice typing for yourself.

On your Android phone. If the Google microphone icon appears in a box where you need to type, tap the microphone. Follow the instructions that appear. Otherwise, you should see the microphone icon on the keyboard. Tap the icon key, and start talking after the beep.

If neither of these options is available, swipe down from the top of the screen to see your Notification panel, and tap Select keyboard. Tap the circle for Google voice typing.

On your Android tablet. Find the Multifunction key to the left of your keyboard space bar.

1. Tap and hold it to make options appear.
2. Tap the microphone to turn on voice typing.
3. Start talking. If nothing happens, check the screen. If you see *Tap to speak*, tap there, and try again.

On your iPad or iPhone. Tap the microphone key on your keyboard, and start talking. When you finish, tap *Done*.

If tapping the microphone key triggers a dialog box that asks whether to enable dictation, tap *Enable dictation*. Or tap *Learn More* if you need help deciding whether to set up voice typing.

Become the fastest texter in town

Are you still sending texts using the old "hunt and peck" method? There's a faster way. Stop typing and start "swyping" for faster, more accurate texting and typing.

Ace Swyping 101. Swyping is like playing connect-the-dots. But instead of using a pencil to connect dots, you use your finger to connect the letters that spell a word. This brings new meaning to the phrase "without lifting a finger."

To swype the word "type," place your finger on the letter t and quickly — without lifting your finger — draw a line through y, p, and e respectively. It's typing by swiping through the letters, and experienced swypers say you'll be amazed at how much faster you can text once you get used to it.

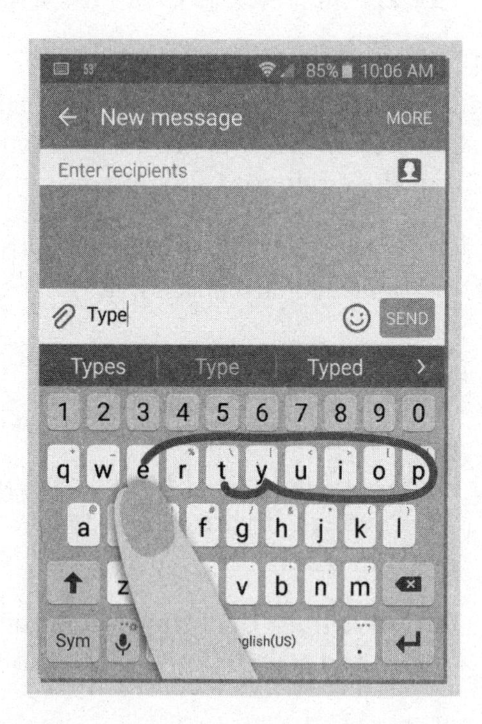

While this typing option isn't available on every phone and tablet, many Android devices come with it preinstalled. If you have an iPhone, you'll need to download an app like SwiftKey or Swype.

Secrets to swyping like a pro. Swyping may seem a little tricky at first, so you may need to practice. You'll know you're swyping correctly when a line forms between the letters as you swype.

Some people say it's easier if you don't try to sweep through the center of every key as you swipe. You only have to "land" on some portion of the key as shown in the example.

Need to type a double letter like in call, oodles, or add? Swype the letter, and then make a tiny scribble or loop to type the letter again. If you keep hitting the wrong key by accident, try turning your phone sideways to type more easily and accurately.

If swyping makes alternate keys appear above the regular keys, just swipe through the letters faster. It may be a challenge, but it should solve the problem.

Help! It's not working. Don't worry. You may just need to turn on a setting. To do this on your Galaxy S6:

1. Tap Apps > Settings.
2. Under the *System* header, tap Language and input > Samsung keyboard > Keyboard swipe.
3. Tap the circle for *SwiftKey Flow.*

It's easy on your Galaxy S tablet, too.

1. Tap Settings > General > Language and input.
2. Tap Samsung keyboard > Keyboard swipe.
3. Tap the circle for *Continuous input.*

Show autocorrect who's boss

"No, it's Missy, not Mossy!" It can be so frustrating when autocorrect jumps in with the wrong word after you typed it right. But don't fret. You can reject changes you don't like. And if autocorrect gets it wrong more than right, you can even turn the silly thing off.

On your Android phone or tablet. Autocorrect comes with brakes, but if you blink you'll miss your chance to use them. As you type, notice the suggestion bar above your keyboard. Just before autocorrect gleefully replaces what you type, it displays a check mark on the left side of that suggestion bar.

That's your only chance to say no. The moment you hit the space bar, autocorrect will replace the word you just typed. But if you tap that check mark *before* the space bar, autocorrect leaves your spelling alone, and the word stays exactly as you typed it. (It has never been proven that autocorrect sulks after you do this, but it hasn't been disproven either.)

If you'd rather skip all the fuss and turn autocorrect off, it's as easy as 1-2-3.

1. The first step depends on whether you're on your phone or tablet.
   - On the Galaxy S6, tap Apps > Settings > Language and input.
   - On the Galaxy Tab S, tap Settings > General > Language and Input.
2. Tap *Samsung keyboard.*

3. On your tablet, slide the Auto replace switch left to turn it off. On your phone, tap *Auto replace*, then turn the slider off.

On your iPhone or iPad. Type a word the iPhone or iPad doesn't recognize, and you may see a bubble above the word, offering a correction. If you have predictive text on, you'll also see a suggestion bar above your keyboard. Choose one of these:

- To keep your original spelling, tap the X in the bubble to delete the suggested replacement. If there's no bubble, tap the suggested word that matches it.
- To make a change, tap the word you want — either in the bubble or in the suggestion bar.

To turn off autocorrect so you can get some peace, tap Settings > General > Keyboard. Turn off the slider for *Auto-Correction*.

Cut, copy, or paste like a pro

Smartphones and tablets don't have a mouse, so how can you select text to move or copy? Let your fingers do the walking! Simply tap the text to move or copy it to different spots in the same app — or to an entirely different app. Just follow these easy steps.

On your Android phone. These instructions for the Galaxy S6 may also work for other Android phones.

1. Press and hold directly on one of the words you want to select.
2. When the start and end handles appear, drag each one to select the text you want.

3. A menu appears either nearby or in the upper-right corner. It offers options to *Select All, Cut, Copy*, and *Paste*. Some apps may only display icons with no text explaining what they mean. If you aren't sure which icon is the one you want, use the guide below.

4. If you want to select all available text, tap the *Select All* icon.

5. Tap the *Cut* or *Copy* option.

6. Navigate to where you want to paste your text, either in this app or another app.

7. Press and hold that spot, tap *Paste*, and voilà, your text appears.

On your Android tablet. Selecting text on your tablet is the same as on your phone, but cutting and copying are a little different.

1. Press and hold one of the words you want to select.

2. When the start and end handles appear, drag them to select the specific text you want, or tap *Select All* in the upper-right corner to select everything.

3. In the upper-right corner, tap *Copy* or *Cut*.

4. Navigate to where you'd like to paste the text — either in this app or another app — and press and hold that spot.

5. Tap *Paste*, and your text appears.

On your iPhone or iPad. These steps are just as easy as the Android's.

1. Press and hold one of the words you want to select. In some apps, you may need to double tap the word instead.
2. A menu appears above the word. Tap *Select* to select part of the text, or tap *Select All* to select all of it.
3. If you tapped *Select*, drag the blue start and end handles to select the text you want.
4. If the menu is not visible when you're done, tap the selection to make it appear again.
5. Tap *Cut* or *Copy*.
6. Navigate to where you want to paste the text. You can paste either in this app or another app.
7. Press and hold that spot. Tap *Paste*, and your text pops in.

Keyboard hard to type on? Split it!

Typing on a tablet keyboard can be frustrating, especially if you're writing a long email. You may feel like you're all thumbs trying to hit the right keys. But you can make your keyboard more user-friendly with one simple trick. Just split it in half.

Dividing the keyboard on the iPad couldn't be easier. Whenever it appears on the screen, put your thumbs on either side, and spread your hands apart. The keyboard splits in half and even shrinks a little to make it easier to reach.

Push the sides back together to restore the keyboard. You can also press and hold the keyboard icon in the lower-right corner, and then select *Split.*

It takes a few more steps to get to the split keyboard setting in the Samsung Galaxy Tab S. Press and hold the microphone key to the left of the space bar. You'll see several options, but choose the keyboard icon at the end of the row. Then you have three choices in the menu.

- *Normal* – the regular layout
- *Floating* – an even smaller keyboard you can drag around on the screen
- *Split* – splits the keyboard in two halves with one end on each side of the screen.

After tapping *Split*, you'll get the smaller, divided keyboard. To return to normal, press and hold the microphone key again, tap the keyboard icon, and then tap *Normal* when the menu appears.

Shake it up to undo typing

"Shake it up, baby now. Shake it up, baby. Twist and shout!"

That's how one of the Beatles' greatest hits begins, and those famous lyrics can help you remember a faster, easier way to trigger Undo on your iPhone or iPad. It's called "shake to undo."

To use this time saver, gently shake your iPhone or iPad the same way you'd shake a box of cereal to hear how much is left. When an

Undo Typing dialog box pops up, tap *Undo* to delete your last entry, or tap *Cancel* to get back to work without undoing anything. It's just that simple.

Take a snapshot of your screen

Want to prove you reached almighty master guru level on your favorite smartphone game — or capture a crazy error message or fascinating text conversation? Take a screenshot. Like any picture, it really will last longer. And it's so easy to do.

On your Android phone or tablet. First, find the right buttons.

- On the Galaxy S6, the Home button is on the front of the phone near the bottom. You'll also need the Power button on the phone's side in the middle of the rim.
- On your Galaxy Tab S, find the Home button on the bottom between the screen and the rim and the Power button on the rim near a corner.

Press and hold both buttons at the same time. The screen edges will flash white while the camera sound clicks.

Finding your screenshot. To see your masterpiece, swipe down from the top edge of the screen to view your Notifications. Tap the one that reads *Screenshot captured.* The next steps depend on which device you use.

- On your phone, tap *Gallery* in the *Open With* dialog box to see your screenshot.

- On your tablet, tap the thumbnail of the screenshot you want. (If you see no thumbnails, tap the screen to display them across the bottom.) Choose from options in the Toolbar along the top of the screen to determine what to do with your screenshot. (Tap the screen if you don't see options.)

To retrieve an Android screenshot later on, tap Apps > Gallery. Your screenshots will be waiting for you there in all their glory.

On your iPhone. Find the Home button below the iPhone 6 screen. Then find the Sleep (On/Off) button on the phone's right rim. Press and hold both buttons at the same time, then let go. You'll see the screen flash white and hear the camera click as it takes the shot. The screenshot will be stored in your Photos app.

On your iPad. The Sleep button on an iPad Air is located at the top of the device instead of along the side. But you still press and hold the Home and Sleep buttons at the same time, then release to take the screenshot. Easy as pie!

Get twice as much done with a split screen

Say you're using your phone for navigating a car trip, but you need to text your friend when you'll get there. Now you can do both at the same time.

Your Galaxy S6 phone, Galaxy Tab S tablet, and iPad Air 2 can split the screen between two apps, so you can use both at once. You'll be surprised at how easy this is.

On your Android phone or tablet. Android calls this feature Multi window. Here's how to use it.

- On your phone: Press and hold the Recent Apps button below your screen at the left.
- On your tablet: Press and hold the Back button on the bottom right, or swipe from the right edge of the screen toward the center.

If a message pops up, tap *OK* to continue. A series of app icons will appear at the bottom of your phone screen or along one side of your tablet screen. Scroll through the list to see all the apps you can use in split-screen mode.

Open the first app you want by tapping its icon. Scroll to the second app, and open it. Each app opens in its own half of the screen. Now you can use both apps at the same time.

If you want to change the amount of screen real estate each app takes up, drag the border between the apps up or down (on your phone) or left or right (on your tablet.)

To reap the full advantages of split-screen apps, tap the circle on that border. A toolbar appears with icons representing these choices.

- Make the apps switch places on your screen.
- Drag and drop content from one app to another.
- Minimize or maximize an app.
- Close an app.

On your iPad. Tap the first app you want to open. For now, that app fills the whole screen. To open the second app, swipe from the right edge of your screen toward the center.

This opens a listing of apps that can be used in split-screen view. Tap the one you want, and it opens on the right side of your screen. (If an app opens there before you can choose, just swipe down from the top of that app, and the list should appear.)

Once you tap the second app you want, both apps should appear side by side with a border between them. If both apps are equipped for split-screen view, a handle appears in that border.

Slide the border between the two apps until each app uses about half the screen. Now you can use both apps and even drag and drop text or images from one to the other.

When you no longer want split-screen view, slide the dividing border to the left or right side of the screen to make one of the apps fill the screen — or tap the Home key to exit.

Make life easier with a personal assistant

Stop scrolling through all those apps on your smartphone just to find the ones you use most. Use your phone's digital personal assistant, and find what you're looking for fast.

Instant answers to life's daily questions. Just ask Siri. If you have an iPhone or iPad, you may have heard of this wonder woman who magically answers questions and does anything you ask.

Her Android counterpart is called Google Now. They know everything because they use the Internet to get the answers you need. And they do it fast. You can ask them questions like:

- what's the weather forecast for tomorrow?
- who won the Cubs game?
- how much is a 15 percent tip on a $35 bill?
- how many ounces in a one-liter drink? You can also try a command like "Convert one liter to ounces."

Siri or Google Now will either answer your question out loud or show you the appropriate Web search results.

Let your assistant handle everyday tasks. Do you hate typing in calendar events and reminders? Tired of fumbling to find your contact list when you want to make a phone call? Your digital assistant can do these for you, freeing up time for more important tasks. Just give it a command:

- Create calendar event, Martha's birthday lunch, November 1, at 1 p.m.
- Remind me to pick up the birthday cake on Friday at 5 p.m.
- Call Osgood at work.

You can even ask them to open apps, set an alarm or timer, ask for directions, and send emails or texts.

So how do you make this magic happen? Simply press and hold the Home button on your iPhone or iPad to make Siri listen for your commands. It's even easier on an Android phone or tablet. Say "OK, Google," and listen for the beep.

It might feel weird to talk to your phone, but give it a try. You'll love what your virtual personal assistant can do for you.

Just remember, Google and Apple may use these clever assistants to collect a lot of information about you for marketing and advertising. This information is drawn from your commands and searches, and possibly from your emails and contact lists.

If you'd rather not give that information away, these digital assistants may not be right for you.

Get a new, portable scanner for free

Finally get rid of all those paper piles! The only tools you need are your tablet or phone, a scanner app, and a place to store and view your scans. Here's how you start.

Gather your scanning tools. That great high-resolution camera on your phone or tablet is perfect for photographing business cards, receipts, and many other documents. All you need is an app to help turn your device into a scanner.

When choosing a scanner app, look for one that's free. If the app includes free OCR text recognition, even better. That will allow you to do text searches of your scans so you can find things fast.

You'll need a place to store your scanned documents. If you use a cloud service like Dropbox or iCloud, check whether the app offers access to these services.

If not, make sure you can either email scans to yourself or transfer scans from your phone to your computer using your wireless network or a USB connection.

Download your scanner app from Google Play for Android devices or iTunes for iPads and iPhones. Check the permissions an app requires before you download.

Use this handy getting-started guide. Each scanner app works a little differently, but these sample instructions show roughly how you'd use one.

1. Tap the app icon to open the app. You may be asked to register an account using an email address and password, or you may see an option to skip that step.
2. To scan, click the scanning icon, and aim your phone camera at the document. Check for three things before you pull the trigger.
    - Make sure the document is flat.
    - Frame the entire document in your viewfinder.
    - Check that you can read all the text through the viewfinder.
3. Tap the icon to scan your document.
4. The scanner app may automatically detect the edges of the scannable image, but you may need to manually make corrections with tools the app provides. You may also see options to rotate, improve, or correct the image.
5. After the first page is scanned, you may see options to add more pages, or you may have to scan one page at a time.
6. When all pages are scanned, transfer the files to storage.

Be smart about storing files. The app may automatically offer limited storage on its own cloud service. But watch out for added charges if you go over your storage limit. To keep your scans and avoid paying the app maker for more storage, use the features that help store your scans on your computer or favorite free cloud service.

Remember that uploading scans of large documents, or many scans throughout the month, will use a lot of cellular data. So you may want to wait to transfer them to storage until you connect to your wireless network.

Once you've transferred your scan, consider deleting it from your phone or tablet, particularly if it contains sensitive personal or financial information.

Hey smartphone, remind me when I get there

Forgetting to buy milk at the supermarket is one of the top 25 things people forget, a British survey reports. If that sounds like you, let your cellphone help remind you. Just tell it what you need, and it will alert you when you arrive at a certain location — like the grocery store. How cool is that?

Location-based reminders are available from three sources:

- the Reminders app that comes with iOS 9 on your iPhone.
- Google Now reminders on your Android phone.
- third-party apps that offer location-based reminders.

On your iPhone. Use Siri to set a location-based reminder for home or work. First, ask Siri, "Where do I live" or "Where do I work" to make sure she has the correct information. Then set a reminder by saying something like, "Remind me to feed the cat when I get home."

If you're in a place where you can't talk, set the reminder silently instead.

1. Open the Reminders app. Tap an empty space, and type your reminder.
2. Tap the i or information icon, and turn on the *Remind me at location* slider.
3. Tap *Location*. If you're prompted to allow the *Reminders* app to use location services, tap *Allow*.
4. A list of locations appears. Select the one you want, or add a new location by typing it in the search bar at the top of the screen, then tapping the location in the list.
5. Tap either *When I Arrive* or *When I Leave* to set when you receive the reminder notification.
6. On the map that appears, pinch or expand the blue circle to set the size of the location. If the circle won't shrink or grow, make sure one of your fingers is on the black dot, and try again.
7. Tap *Details*, and then tap *Done*.

If you have problems with location-based reminders, check whether Location Services are on. Tap Settings > Privacy > Location Services, and turn on the *Location Services* slider. But don't stop there. Scroll down to the *Reminders* app, tap it, and then tap *While Using the App*.

On your Android phone. If you already use Google Now, setting location-based reminders is easy. (For more information on

Google Now, see page 189.) Start by saying "OK Google." Listen for the beep, say "Set reminder," and tell Google what you need.

If you're in a place where you can't speak aloud, type "set reminder" or "remind me to" in the Google search box and tap the magnifying glass.

1. On the *Add a Reminder* screen, tap *Title*, and type your reminder.
2. Tap the circle next to *Place*.
3. Check the location that appears. If it's the one you want, select it. Otherwise, add a location by tapping the drop-down arrow at the right. Select a location, or tap *Set location*. If you picked *Set location*, type the name or address of the place you want. When it appears in the list, select it.
4. Tap the check mark to save your reminder, or the X to delete it.

If location-based reminders in Google Now don't work properly, tap Apps > Settings > Privacy and safety > Location, and turn the *Off* slider on.

Take advantage of helpful apps. If you already use a productivity app such as EasilyDo or AnyDo, check whether it offers location-based reminders.

Decide when your phone should ring and ding

Suppose you're a secret agent out to save the world. Since your mission depends on being stealthy, your smartphone can't make any noise — unless your boss calls or messages to warn of impending danger. Thankfully, your phone's Do Not Disturb mode can take care of that problem and save you from possible disaster.

Enjoy the sounds of silence on your Android phone. Tap Apps > Settings > Sounds and notifications > Do not disturb. Turn the slider on. Tap *Allow exceptions*, and turn on the sliders for the exceptions you want.

- *Alarms.* This setting allows alert sounds for any alarms you've set up.
- *Calls.* This allows all incoming calls to ring — not exactly good for staying stealthy.
- *Messages.* All messages will trigger a notification sound.
- *Calls and/or messages from.* This setting limits which calls ring and which messages trigger a notification sound. Choices include *All, Favorite contacts only*, or *Contacts Only.* Make your boss a favorite contact, select that group, and you won't miss his warning call.
- *Events and reminders.* Is saving the world on your calendar? It will remind you that today's the day unless you turn this setting off.

Program peace and quiet on your iPhone. Tap Settings > Do Not Disturb. Turn on the slider labeled *Manual.* Now all calls and alerts will be silent. If you want it to stay that way, scroll down to the *Silence* option and make sure *Always* is checked. Otherwise, it will only stay silent when the phone is locked.

If you want to allow calls from anyone — like your boss — tap *Allow Calls From* and make your choice:

- *Everyone.* Selecting this allows all calls to ring.
- *No One.* This keeps every call silent and sends each call to voicemail.

- *Favorites.* Only calls from people you've picked as favorites will ring.
- *Groups.* Calls from anyone in your designated group will ring. Other calls will make no sound and roll to voicemail.

If you need a quick way to mute all calls and alerts, swipe up from the bottom of your screen and tap the moon-shaped icon. This instantly puts your iPhone into Do Not Disturb mode.

You can do the same on your Android phone by swiping down to open the Quick Settings panel and tapping the Do Not Disturb icon.

Mission accomplished? You can now turn off Do Not Disturb mode on either phone by going back in and turning the slider off or by retapping the icons.

Silence the howl of emergency alerts

Emergency alerts about dangerous weather and other disasters could save your life. But for people who own two cellphones, hearing that howling siren coming from both phones at once can be extra jarring. Here's how to turn off emergency alerts on at least one phone.

On your iPhone. Tap Settings > Notifications, and scroll down to the *Government Alerts* section. To silence *AMBER Alerts* and *Emergency Alerts*, tap the button next to each.

On your Android phone. These settings occur in different places on different Android phones. On some Galaxy S6 phones, the settings are buried in the Messages app. Tap Apps > Messages and follow these steps.

1. Tap *More* in the upper right corner.
2. Tap Settings > Emergency alert settings.
3. Tap *Emergency alerts.*
4. To stop receiving alerts for natural disasters and dangerous weather, tap to turn off *Extreme alert* and *Severe alert.*
5. To stop receiving alerts for missing children, tap the button beside *AMBER Alerts.*
6. Tap the button beside *Emergency alert test messages* to end those as well.

You cannot turn off Presidential alerts — even if you didn't vote for that person.

If you don't find Emergency alert settings in your messaging app, look for an app named Emergency Alerts. That's where some Android phones, including some Galaxy S6 models, hide these settings. Still no luck? Look for them in the Settings app.

My phone still howls out alerts. The instructions above only turn off the special WEA alerts issued by the government. If your phone came with weather apps, or if you've installed any weather or disaster apps, you may need to change the settings in those apps.

Find your missing phone in minutes

Oh no, you lost your phone. Or worse, someone stole it. Never fear — here's how to locate it fast.

Set yourself up for success. Before you lose your iPhone, turn on its locator beacon. Tap Settings > iCloud > Find My Phone. Slide the *Find My Phone* toggle to the right to turn it on.

You can't find a lost Android phone unless the right Google and phone location settings are turned on. Here's how to do it.

1. Tap the Apps icon, then tap the gear-shaped Google Settings icon. You may need to look in the Google folder to find it.

2. Scroll past *Sign-in & Security*, and tap *Security*.

3. Under *Android Device Manager*, turn on the slider for *Remotely locate this device*.

4. Turn on the slider for *Allow remote lock and erase*.

Next, make sure the phone's location settings are on. Swipe down from the top to open your Notifications, and tap the Settings icon. Tap Privacy and Safety > Location, and turn the slider on.

Pinpoint your iPhone. If your iPhone ever vanishes, open a Web browser, and follow these steps.

1. Type *icloud.com* in your address bar, and go to the site.

2. Log in to your iCloud account, and click the *Find My iPhone* icon.

3. At the top of the screen, click *My Devices* to trigger a drop-down list of your Apple devices. Click the one you want, and your phone's approximate location will appear on the map.

4. If the phone is within hearing distance, click *Play Sound* in the nearby dialog box. If you hear a high tone, find your phone, and unlock it.

5. A *Find My Phone Alert* notification may appear on your screen. Click *OK* to dismiss it.

Locate a missing Android phone. Android Device Manager can help you track down your phone. Using any Web browser, type *android.com/devicemanager* in the address bar. Sign in with the name and password of the Google account registered to your Android phone.

1. A map appears with a dialog box in the upper left corner. This displays your phone's name. If you have more than one phone on this account, tap the nearby arrow to switch to the correct phone. Finding the phone may take a few minutes.
2. When the phone is located, a circle appears on the map showing roughly where the phone is. This may also trigger a notification on the phone.
3. If the phone is within hearing distance, click *Ring* to make the phone ring loudly for five minutes. Press the Power button to stop the ringing.

If you get the *Location Unavailable* message, Device Manager may not be able to find your phone because:

- your phone isn't connected to Wi-Fi or cell service.
- the phone isn't signed in to your Google account, or its Location services aren't turned on.

When your iPhone or Android is out of reach. Two other options are available from either Find My iPhone or Android Device Manager.

- *Lost Mode* (iPhone) or *Lock* (Android) allows you to lock the phone from miles away so anyone who finds it has to type the passcode or pin to use the phone. You may also have the option to put a contact number on the lock screen.
- *Erase iPhone* or *Erase* (for Android phones) erases your phone's memory and all its data, including personal information that could be used for identity theft.

Never lose your phone contacts again

Whether a phone disaster erases all your contacts, or your phone is lost or stolen, you don't want to spend tedious hours rebuilding that list from scratch. So do this instead — and save yourself a lot of hassle.

Keep an updated copy of your iPhone contacts. You can back up your contacts to the cloud daily if you signed up for an iCloud account. But you may have to turn on a setting first. To do this, Tap Settings > iCloud, and scroll down to make sure the slider for *Contacts* is turned on.

Now your contacts will back up automatically when your device is turned on, locked, and plugged into a power source. It also must be connected to Wi-Fi.

Back up Android contacts auto-magically. Most Android users have a Google or Gmail account connected to the phone. To check whether you do, tap Apps > Settings > Accounts. Tap Google or Gmail to see the name of the account. Tap the Gmail account name, and turn on the *Sync contacts* slider.

When you're ready to back up your contacts to your Google account, open your Settings, and follow these steps.

1. Tap *Backup and reset.*
2. If the *Back up my data* slider is not turned on, turn it on now.
3. Check whether you see a Gmail account listed below *Backup account.* If not, tap *Backup account* and add your Gmail address.

Now your contacts will be backed up to the Google cloud automatically.

If you want to see whether the backup is working, sign in to your Google account on a PC or Mac. In the upper-left corner, click on the drop-down arrow beside the word Gmail, then click *Contacts*. All your contacts should appear.

If you don't see all of them, check the menu along the side of the page, and click *All contacts*. If they still don't appear, your contacts may be backing up somewhere else.

Check for backup apps. Google and iCloud may not be the go-to accounts your phone backs up to. Your wireless carrier may have pre-loaded an app to handle the process, or you may have downloaded an app from your carrier, iTunes, or Google Play to back up your contacts for you.

If your contacts are not backing up properly to iCloud or Google, check for backup apps on your phone, or contact your wireless carrier for help.

7 ways to keep hackers and spies out of your smartphone

Do you realize your cellphone can be "hacked," in some cases more easily than your home computer? It's true. Here's how to protect your privacy, your security, even your bank account from malicious phone pirates.

Lock the door. You wouldn't leave your house or car unlocked. Don't leave your phone unlocked either. That could make it easier to hack than your home computer.

Always put a password, pin, or passcode on your lock screen, so no one can look up personal information on your phone when your back is turned.

Avoid the threats of public Wi-Fi. If you must use public wireless networks in places like airports and coffee shops, avoid using email, making online purchases, or handling anything financial. These unprotected networks are easy for hackers to hijack.

Sometimes, hackers even trick you into logging in to their network by making it look just like the network you thought you were using.

Protect your password security. Don't save the password for any apps or Web services you use on your phone, especially if you use a banking app or other app with financial information.

Keep malware and spyware out. Only download apps from Google Play, iTunes, and other official app stores. Unofficial app stores are more likely to carry apps with malware hidden inside.

Read app permissions before you accept them and install the app. If you see a permission you don't immediately understand, don't download the app until you can get more information.

Also, hold off on downloading an app if you have any concerns about the information it collects. Spyware can be hidden on some phone apps.

Frustrate hackers regularly. Install security updates and patches as soon as they're available.

Thwart phone thieves from accessing data. Set up Find my iPhone on your iPhone or Android Device Manager on your Android phone. These apps can help you find a lost phone or erase your data from a stolen phone.

For more information on using these apps see *Find your missing phone in minutes* in this chapter.

Install a security app. Shield your phone with free security software from companies like Avast.

Stop thieves from using your stolen phone

An IMEI number helped reunite a journalist and his lost phone in Nepal, and it could help do the same for you, too. So what on earth is an IMEI? Your social security number works as your identification, right?

Well, your phone has numbers that serve as its ID. Depending on which phone you have, this may be an International Mobile Equipment Identifier (IMEI), a Mobile Equipment Identifier (MEID), or both.

It only takes a few minutes to find these numbers, either in your Settings app or inside the battery compartment, so do it now. For phones with a removable battery, pop out the battery and check the label inside the compartment. For phones without removable batteries, follow these steps.

- On your Galaxy S6, tap Settings > About phone > Status. Next, tap *IMEI information* to see the IMEI and MEID numbers, or look for the IMEI number in the list of stats.
- On your iPhone, tap Settings > General > About, and look for IMEI or MEID.

Copy the numbers and hide them in a safe place, in case your phone gets lost or stolen. They could hold the key to helping the police return it. They can also help your cellphone carrier lock down your device if it falls into a thief's hands.

Immediately report the theft or loss to your carrier and provide your IMEI or MEID. The company may be able to disable the phone, so that a thief can't run up your bill with fraudulent charges.

No-fuss way to deposit checks

Remember when you actually had to go into a bank to deposit a check? Not anymore. With the right app, you can simply snap a photo of the check with your phone and send it off electronically to your account. No fuss, no bother.

If you're ready to join this amazing world of high-tech banking, you'll first need to download your bank's specific app from an authorized app store such as Google Play or iTunes.

Then, before you use mobile deposit, contact your bank and ask these questions.

- How should I endorse the check? Some banks have special instructions for this. For example, you may be told to add "For mobile deposit only."
- How long before the money becomes available?
- What happens if fraud or losses occur?
- After remote deposit, what do I do with the paper check, and how long do I keep it? Experts recommend you immediately void the check to prevent anyone from stealing and cashing it (or to keep you from depositing it a second time by mistake!)

When you're ready to take a picture of your check, rest it on a contrasting background, and make sure the entire check can be seen in the frame.

Snap a picture of the front of the check and the back. Review the photo after you take it to be sure you can read everything easily.

Make private files invisible to everyone else

You have certain files on your phone you'd like to keep safe from prying eyes. Lucky for you, Samsung phones and tablets have a feature called Private mode that lets you hide and password-protect photos, videos, music, and documents.

Turn Private mode on, and you'll see the commands that allow you to hide and password-protect your files. After you select your files, turn Private mode off to make them disappear.

Poof! Both the commands and the files you hid will vanish as if they'd never been there. Want to see them again? Just turn Private mode back on.

Turn on your file-hiding system. Here's how to use this handy setting on a Galaxy S6.

1. Tap Apps > Settings.
2. Scroll down to the *Personal* section and tap Privacy and safety > Private mode.
3. Turn the *Off* slider on to activate Private mode.
4. Hidden files can be protected by a password, pin, pattern, or fingerprint ID. Tap the one you prefer.

On a Galaxy Tab S tablet, you can turn on Private mode by tapping Settings > Device > Private mode and following the prompts.

How to hide files. Pick a file to hide on your Galaxy S6 or Galaxy Tab.

1. Tap the app you normally choose to view or use the file you want to hide. Your app choices may include Gallery, Video, Music, and My Files.
2. Tap the file you want to hide.
3. On the Galaxy S6, Tap More > Move to Private. On the Galaxy Tab S, tap the three vertical dots in the upper-right corner, then *Move to Private.* If you don't see *More* or the three dots, tap the screen to make them appear.

If *Move to Private* doesn't appear, tap *Edit* or *Select.* Select the file to hide by tapping its checkbox, and try Step 3 again.

You can hide as many files as you want. If you accidentally hide the wrong one, just open its app, tap the file you want, and tap More > Remove from Private.

Turn off Private mode. Remember, your files don't disappear until you turn Private mode off. So do that now.

- On your Galaxy S6, tap Apps > Settings > Privacy and safety. Then tap *Private mode* and turn off the *On* slider.
- On your Galaxy Tab S, tap Settings > Device > Private mode. Turn off the slider.

Not a Samsung owner? You're not out of options. Other manufacturers may also offer similar tools. For example, some LG phones offer Guest mode and ways to hide pictures, while some Apple devices may allow you to tuck photos away in a hidden directory. Check the online help for your device for more details.

Forgot your password or pin? Sneak in through this back door

Securing your phone with a password is smart, but what if you forget it? No worries — here's how to recover it.

On your Android phone. Your phone has a feature called Smart Lock that helps you bypass your lock screen password or pin under secure conditions.

For example, say you forget your password, but you've set up your home as a trusted location in Smart Lock. When you get home, your phone turns off its pin or password lock. Then you can change your screen password or pin, preferably to something you'll remember forever.

If you already use a pin or password, you can set up Smart Lock. Tap Apps > Settings > Lock screen and Security. Then tap Secure Lock Settings > Smart Lock. Choose one of these Smart Lock settings that will unlock your phone or keep it unlocked until conditions change.

- *Trusted devices.* Your phone turns off its pin or password lock while connected to a trusted device through Bluetooth or Near Field Communication (NFC) technology. You specify which devices are trusted, such as an NFC tag or Bluetooth-connected fitness tracker.

  Be careful about choosing a Bluetooth device your phone stays connected to most of the time. That may be almost as risky as using no pin or password at all.

- *Trusted places.* You specify a secure location, such as your home, where the pin or password lock can stay off. This setting requires an Internet connection, and your phone's Location services need to be turned on.

Just remember that your circle of trust may extend more than 300 feet from your location — almost the length of a football field! If someone else has your phone outside your trusted place, they could easily access it.

- *Trusted voice.* You can unlock the phone by saying "OK Google" if you use Google Now. (For more information on Google Now, see *Make life easier with a personal assistant* earlier in this chapter.)
- *On-body detection.* Turn on this setting, and the phone won't require a pin or password while anyone holds or carries it. The prompt for a pin or password returns after you put the phone down for a minute.

Some phones also offer face recognition as an option.

Using any of these options may raise the risk that an unauthorized person can use your phone, so take extra precautions if you use Smart Lock.

On your iPhone. If you enter the wrong passcode on your iPhone more than six times in a row, you're in trouble. Apple doesn't give you a back door to rescue you from your memory lapse.

Your only option is to erase your phone and restore it from an iCloud backup. Visit *support.apple.com* on your PC or another device to help do this.

If you're lucky enough to be reading this before you forget your password, take steps to make sure you can look it up should you need to.

Jump-start a stalled device

Fancy gadgets get brain freeze, too, a lot like people do. Their screens seize up and they refuse to respond when you tap an icon or press a button. Here's how to give them a nudge and get them moving again.

Take off the kid gloves — literally. Your touch screen may refuse to do anything if you're wearing gloves or have dirty fingers. Lose the gloves, wash your hands, or warm them up. Your screen may start cooperating again.

Play the wait-and-see game. Sometimes the device is only temporarily frozen. Wait up to 15 minutes, and see if the device starts operating normally. That's an especially good approach if you have apps open. If you have to force your device to turn off, you may lose unsaved data or changes in your open apps.

Thaw a frozen Android device. Sometimes a gadget is so frozen that it just won't respond. Never fear. You can generally unfreeze any device by pressing the Power button, although you may lose data in any apps you have open.

- On your Galaxy S6 phone, press and hold the Power button and Volume Down button together for 7 seconds. This will restart — and likely unfreeze — the phone. If a very basic menu appears on your screen, use the Volume buttons to highlight *Normal Boot*, then press the Home button to restart your phone.

- On a Galaxy Tab S, press and hold the Power Lock button. If a menu appears, tap *Restart*. If no menu shows up, press and hold both the Power Lock and Volume Down button for 10 seconds. This forces the tablet to restart.

For other Android devices, check your instruction manual to see if you should press the same buttons as described here. No manual? No problem. You can usually download a copy from the manufacturer's website under headings such as Support.

End the freeze frame on an iPad or iPhone. Always try to close your apps before following these steps. Otherwise, you'll lose any data you haven't saved.

Press and hold the Power button for several seconds. The message *Slide to power off* should appear on the screen. Swipe right and see if the device shuts off. If it does, press and hold the Power button to restart your device.

If it doesn't turn off, press both the Power and Home buttons simultaneously. Keep pressing them for a solid 10 seconds, even if the screen goes black or the *Slide to power off* message appears. Only release the buttons when you see the Apple symbol. That means the device is restarting.

Troubleshoot your device like a tech expert

When NASA's New Horizons space probe developed problems on its way to Pluto, NASA fixed things by switching the computer to Safe Mode. You can do the same with your Android phone and tablet.

What can Safe Mode do for you? It can help diagnose problems with your phone or tablet the same way it does on a full-fledged computer. On your laptop or desktop, Safe Mode shuts down all but the essential programs, so you can tell which app, program, or update is causing the problem.

On your Android device, Safe Mode loads the bare essentials — the Android operating system — without any third-party apps or extra software. This helps you figure out whether the Android OS or some downloaded app is to blame for your gadget's troubles.

How to reach Safe Mode. If your phone or tablet begins acting funny or stops responding, and if you suspect a newly installed app is to blame, then try restarting in Safe Mode.

First, turn off your device. On a Galaxy S6 phone or Galaxy Tab S tablet, that means you press and hold the Power button until a menu appears. Tap *Power Off*. If a second menu appears, tap *Power Off* again.

Now you can reboot in Safe Mode. On the Galaxy S6 phone:

1. Press and hold the Volume Down button.
2. At the same time, press the Power button for a few seconds.
3. Release the Power button when the phone starts up, but keep pressing the Volume Down button.
4. When the device finishes restarting, you should see the words *Safe Mode* in the lower-left corner of the screen. Then you can release the Volume Down button.

The steps are the same for the Galaxy Tab S, except for one thing. Instead of pressing the Volume Down button, you press both Volume buttons (Up and Down). You can release them both once you see the words *Safe Mode* on your screen.

Some devices save you all this trouble by offering a *Reboot in Safe Mode* option when you press and hold the Power button. Other phones may imitate the Galaxy Tab S, making you press and hold both Volume buttons (Up and Down), instead of just the one.

Uninstall problem apps. Once you're in Safe Mode, notice whether your gadget still acts strangely or if the problems go away. If they vanish, you'll know a recent update or new app is the likely culprit. If you think a new app is the problem, try uninstalling it and restarting your device.

To uninstall an app, tap Apps > Settings > Applications > Application Manager. Find and tap the app, then tap *Uninstall.* When finished, restart the device as you normally would — usually by pressing and holding the Power button, then tapping *Restart* in the menu that appears. This will take it out of Safe Mode.

Web browsers

Your key to the Internet

Make your favorite Web browser the default

Your computer has a default Web browser. That means when you click a link in an email or some other application, it will open the Internet in one specific browser. On Apple computers, the Web browser defaults to Safari. If it's a PC you're most likely running a version of Windows. The default used to be Internet Explorer, but if you've upgraded to Windows 10, it will be Microsoft Edge.

Those aren't the only two browsers out there, though. In this chapter, you'll also hear about tips that apply to two other widely used Web browsers, Mozilla Firefox and Google Chrome. You can install them on either a Mac or a PC. Download Firefox by going to *mozilla.org* and Chrome by going to *google.com/chrome/ browser*. The install pages will detect your operating system (OS) so you download the correct version.

Which browser is better than the other? It depends on who you ask. But whatever you decide, download it if necessary, then make sure it's your default browser — if you've upgraded your OS recently, this setting might have changed without you even knowing it.

On a PC. Click ⊞ > Settings > System. Under *Default Apps*, see what is showing for *Web browser.* Click on it to change.

On a Mac. Go to System Preferences under . Click *General* and select from the drop-down menu next to *Default web browser.*

You can often change the default within a browser itself. Look for Preferences > Make Default or Set Default.

Change your home sweet home page

Depending on which browser you're using, you'll have a different default start, or home, page when you first open your browser. Chrome, for example, has a home page set to a Google search box. But would you like to customize the first thing you see when you open your Web browser? You can have email, sports scores, or headlines ready to read the second you go online.

Here's how to change your home page on a PC.

Chrome. ≡ (or Alt+f) > Settings > On startup.

Edge. ··· > Settings > Open with.

Firefox. ≡ (or Alt+t) > Options > Startup > Home Page.

Here's how to change your home page on a Mac.

Chrome. Chrome menu bar option > Preferences > Settings > On startup.

Firefox. Firefox menu bar option > Preferences > General > Startup > Home Page.

Safari. Safari menu bar option > Preferences > General > Homepage.

Just FYI, a quick way to get to Preferences on a Mac for Chrome, Firefox, or Safari is to press ⌘+comma.

Pick your default search engine

Much like choosing what to have for dinner — or whether you buy a Mac or a PC — choosing a default search engine is all a matter of taste. Google is a heavy hitter, but it's not the only option on the menu. Here's how to change your default search engine.

Chrome. ≡ (or Alt+f) > Settings > Search > Set which search engine is used when searching from the omnibox. In the drop-down menu, you'll probably see some familiar names. Used to Yahoo! or Bing? Pick your old standby or try something different. On a Mac, you'll start from the Chrome Menu bar option at the top of your screen then choose *Preferences* and follow the same steps.

Edge. ••• > Settings > View advanced settings > Search in the address bar with. Click the down arrow and choose your search engine.

Firefox. Find the small search box on the right side of the toolbar. Click on the magnifying glass to get a drop-down menu of alternative search engines. It's sort of like a search engine buffet. You can click *Change search settings* to set a new default. From the new window that opens up, you can also check and uncheck options from the list of *One-click search engines*. Select *Add more search engines...* near the bottom if you want something different. This could be useful if you need to perform a very specific search, such as shopping for a piece of memorabilia on eBay.

Safari. Safari menu > Preferences > Search > Search engine.

The URL secret that will save you a ton of keystrokes

When websites first became a thing, you had to use the full address. Even in talking about them, you might have said "HTTP, backslash, backslash." Now, your Web browsers can figure out where you need to go just by typing, say, "fca.com" into the address bar.

Even better, you don't always need to add the "dot com" part, because there's a shortcut for that. In most browsers, type in the main name, then simply press Control+Enter.

Shorten long Web addresses

Found something great on a website and want to share the link with a friend? Easy. Just copy and paste it.

Start by going up to the Web address box and selecting it all with your mouse. Then use one of these quick ways to copy it:

- Right-click with your mouse or Option+click.
- Type ⌘+c or Control+c.

Then go to your email message or wherever you want to share the link, and paste it with ⌘+v or Control+v.

Uh-oh. That URL just pasted in as several lines of broken text, including letters, numbers, symbols, and all kinds of gobbledy-gook. Not only does it look messy, but because it's broken up, your friend won't even be able to click on it. So instead of sharing a crazy long URL, use a URL shortening service. These are sites that let you paste in your long URL and exchange it for a really short one that takes you to the same place.

Discovered an Elvis Presley wearable blanket on Amazon that you simply must share around? Use the Google URL Shortener at *Goo.gl* and it goes from this: *http://www.amazon.com/Presley-Comfy-Throw-Blanket-Sleeves/dp/B00FGCLPS8/ref=sr_1_6?ie=UTF8&qid=1443112204&sr=8-6*

To this: *http://goo.gl/PTZIyU*

Bitly.com and *TinyURL* are also popular URL shortening sites, but there are dozens more out there. Some of them have toolbars you can install on your browser so they're easier to use.

Tap the spacebar instead of scrolling

Reading Web pages in your browser can become monotonous — you just keep on scrolling, scrolling, and scrolling. If your mousing hand tires out before your eyes do, use the keyboard instead. Just tap the spacebar each time you want to move down one full screen. And if you want to go back up a screen, press Shift+spacebar.

Discover the hidden 'Yes' and 'No' keys

You're furiously typing away, leaving spirited feedback on your favorite website. Then you have to take your hands off the keyboard and reach for the mouse, simply to click the *Post* button.

Or do you? Discover the secret powers of the Enter key that can save you a lot of time and trouble. In just about any situation, you can hit Enter instead of using your mouse to click OK, Print, Save or other action buttons — usually those with a dark outline around them. It's basically a "Yes" key.

So if you have a "Yes" key, do you have a "No" key, as well? Yes, you do. It's Esc, or escape. Anytime you need to Cancel, Quit, or Close a dialog box or message that might come up as you're typing, just press Esc.

Magnify your Web pages for easier surfing

Does the text on Web pages seem to be getting smaller? It's not just your eyes. Sometimes websites use smaller fonts so they can cram as much information as possible around all the ads. But don't fret. Here are some simple ways you can enlarge Web pages and stop squinting.

In Chrome, Firefox, and Safari on a Mac, go under *View* and choose *Zoom In* to magnify or *Zoom Out* to make the page smaller. Or you can simply press ⌘+ (plus key) or ⌘- (minus key). In Safari and Firefox you can also go to View > Zoom Text Only, which lets you magnify words only, not any images or graphics.

If you don't want to bother with zooming in and out, you can set a minimum size for the text on your pages. While in your browser, press ⌘+comma, or go to the Preferences drop-down menu.

- In Safari, choose *Advanced* and check the *Accessibility* box that reads *Never use font sizes smaller than.* Increase the number to 24.
- In Chrome, scroll down to Show advanced settings... > Web content. Choose *Font size* or *Page zoom*, and pick the size or percentage you want.
- In Firefox, click *Content.* Go to *Fonts & Colors* to change your size, or click the *Advanced* button next to it for options galore.

Keep your options open with tabbed browsing

Have you ever found yourself lost in a sea of open windows when surfing the Internet? Perhaps you didn't know that all web browsers offer tabbed browsing, which is a way to keep your various browser windows open but also organized and easy to navigate.

To create a new tab, just click the plus sign or blank page icon to the right of your open tab. Or use keystrokes. On Windows, it's Ctrl+t. On Macs, use ⌘+t. To switch around between tabs, you could of course just click on them. But if you like to avoid the mouse as much as possible, you can rotate through tabs with keystrokes.

On a Mac. Use Control+tab. You can also use ⌘+1, ⌘+2, and so on, to move from one tab to the next.

On a PC. Press Ctrl+tab or Ctrl+1, Ctrl+2, and so on, to rotate through the tabs. Adding Shift goes backwards through the list (unless you're on Edge).

Want even more quick tabbing tips?

- To visit a new website and open it in a new tab, press Ctrl (Windows) or ⌘ (Mac) while clicking on the link.
- To open all of your bookmarks in a folder in Firefox, go to Bookmarks > (Folder Name) > Open All in Tabs.

Reopen a tab you closed by accident

It's happened to all of us. You're multitasking away, switching between screens and programs, when you accidentally close one of your browser tabs. Don't worry. Just do this to get it back.

- On a PC, press Ctrl+Shift+t.
- On a Mac using Firefox or Chrome, press ⌘+Shift+t.
- On a Mac using Safari, press ⌘+z.

Organize your tabs by pinning them

You've gone down the rabbit hole, chasing one interesting topic after another in the vastness of the Web. Suddenly you realize that you have way too many tabs open in your browser, and you can't even figure out which one to go back to so you can check your email. Here's a trick, called pinning, that will help you know what's what again.

Pin a tab and you'll keep it open and active, but shrink the "tab" part to show only a small icon for the website instead of its full name. And you can't accidentally close it, since to do that you have to unpin the tab first.

In Chrome, Firefox, and Safari, right-click on a tab to view the drop-down menu with *Pin Tab* as an option. Once you've pinned a few tabs, you can click and drag to reposition them. To unpin or close the tab, just right-click again and choose your option from the drop-down menu.

Go to a favorite website instantly

Crazy for Craigslist? Wild about Wikipedia? A fanatic for Facebook?

Your favorite websites are probably bookmarked, but wouldn't you love to go to them without even bothering to open your browser first? You can place a shortcut to your most frequented sites right on your desktop. One click and you go straight to your happy place.

In Windows 10, use the Edge browser to open the site. Then go into ··· > Pin to Start. Your site will appear as a tile when you click ⊞.

On a Mac using Safari, click on the website's URL inside Safari's address bar, and an icon representing the site will pop up. Drag that icon down to the dock and place it to the right of the separator near your Trash Can and a blue globe will appear. Mouse over it to see the title of the site. Just so you know, the left side of the separator is for apps, the right side is for URLs and folders.

On a Mac using Chrome or Firefox, click on the URL and just drag the whole thing to the same spot on your dock.

Get Web updates on your desktop

Constantly checking the Internet for email or the latest breaking news? Let your browser send them to you instead. Known as push notifications, these updates appear on your screen from any website you allow to send them.

If you're a Windows 10 computer user, you can easily get updates in your Action Center. (Learn more on page 79.) Mac users, read on.

Not all websites or browsers support notifications. But if a website does, a window will pop up while you're browsing and ask if you want to allow them. Once you say yes, you won't even need your browser open to get the alert.

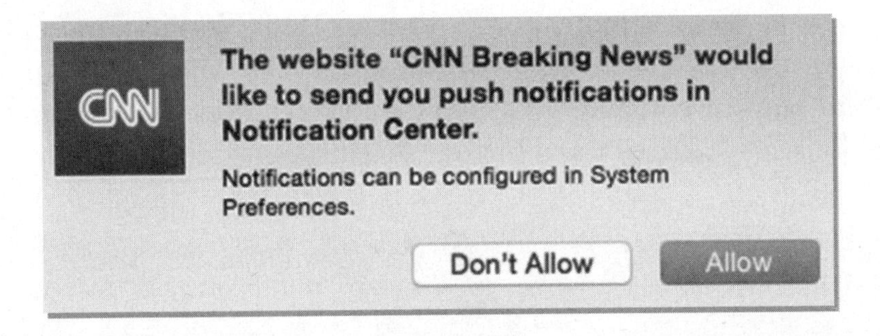

To get email notifications, you do need to be online. For example, if you're logged in to Chrome and surfing the Web, you'll get a desktop notification when a new email arrives in your Gmail account. To set it up, click the gear icon and go to Settings > General > Desktop Notifications. Select your preference under *Click here to enable desktop notifications for Gmail.*

Ready for your daily news fix? CNN.com and Foxnews.com are examples of sites that will send up-to-the-minute news alerts. Here's how to set up your browser to get these and other push notifications.

Safari. Go to the drop-down menu under Safari > Preferences (or press ⌘+comma) and click *Notifications.* There you can check the box *Allow websites to ask for permission to send push notifications.* Click *Notifications Preferences* to manage settings.

Chrome. Click ≡ > Settings > Show advanced settings. Under *Privacy*, click *Content settings*. Go to *Notifications* and select your preferred setting. While in Chrome, you can also turn notifications on and off by looking for the Alert icon (shaped like a bell) at the top right of your screen. Click on the bell, then the gear icon, and uncheck anything you don't want to see.

Firefox. The default is for websites to ask before sending notifications. When you're on a site that offers them, set your preferences by pressing ⌘+i > Permissions > Receive Push Notifications and clicking the button for *Allow*. To adjust how and when notifications appear on your computer, go to System Preferences > Notifications and look for the site or app in your Notification Center.

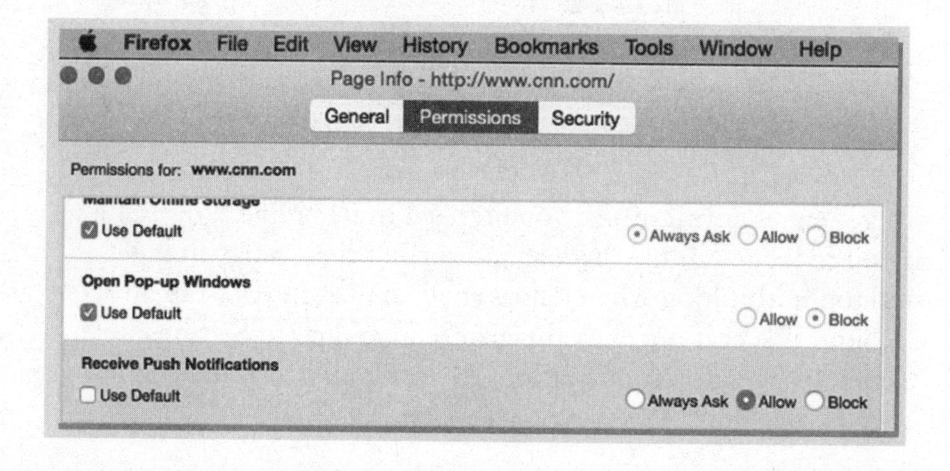

Find bookmarks in a flash

Bookmarking a website for later is kind of like clipping recipes. Before you know it, you have so many that you're overwhelmed just trying to find the chili recipe your cousin sent you last month. That is, unless you can search through them quickly and easily.

Firefox. Click the Bookmark button on the upper right. (It looks like a stack of books.) Then press Control plus the folder you want to sort. Choose *Sort By Name* from the drop-down menu, and the bookmarks within the folder will sort alphabetically.

If your bookmarks are unsorted or you're not sure which folder to search in, click the Bookmark button and scroll down to *Show All Bookmarks*. Click on the three lines icon in the Toolbar, go to *Sort*, and choose how you'd like to sort the bookmarks.

Chrome. Click Option+Command+b on a Mac or Ctrl+Shift+o on a PC to open the Bookmark Manager. Use the search box on the right to find your bookmark. Or you can click on a folder, go to the *Organize* drop-down menu, and choose *Reorder by Title* to put the bookmarks in alphabetical order.

Safari. Click the *Bookmarks* drop-down in the Toolbar and select *Edit Bookmarks*, or press Option+Command+b. Then use the search box at bottom right. Or press Shift+Command+/ and search Safari's help box, which will look in your search history and your bookmarks.

Save must-read articles with the click of a button

That article about the latest cooking craze looks fascinating, but you just don't have time to read it now. Should you bookmark it? Try to remember it for later? Turns out that saving long Web articles or posts for convenient off-line reading later is as simple as a few clicks or keystrokes.

Make a reading list. If you have a Mac and use Safari, check out the Reading List feature. When you're viewing a story that you want to read later, press Shift+⌘+d or go to the Bookmarks drop-down menu and select *Add to Reading List.* This is different from a bookmark because the article is saved in its entirety, and you can read it without having a connection to the Internet.

Total Mac addict? Your Reading List will sync to iCloud so you can also access it on your iPad or iPhone. To access it, look for the open book and then the glasses icon in the Safari Toolbar.

Microsoft Edge offers a similar option called Reading View. It loads a story and strips out all the ads so you only see the text. Just click the open book icon on the right side of the tool box, or press Ctrl+Shift+R. Click to add to your Reading List.

Try a browser extension or bookmarklet. If you use both Safari and Firefox, the latter will import your Safari Reading List. Firefox also has a Reading View, which prompts you to download an extension called Pocket that's also available for Chrome and Safari.

Pocket lets you save online articles, videos, and images, and tag them to make them easier to search for later. After it's installed, you'll just click the Pocket icon in your Toolbar to save. For Edge users, Pocket is available as a bookmarklet, a small piece of software stored on your browser that will appear in your Favorites or Toolbar. Sign up at *getpocket.com*.

Other read-it-later extensions and add-ons available for Safari, Firefox, and Chrome include Instapaper (*instapaper.com*) and Readability (*readability.com*). Instapaper turns articles into black and white newspaper-type pieces that you save as bookmarklets, email to yourself, or send to a tablet. Readability doesn't change the color, but it works in much the same way so you end up with a clutter-free article to read whenever you're ready.

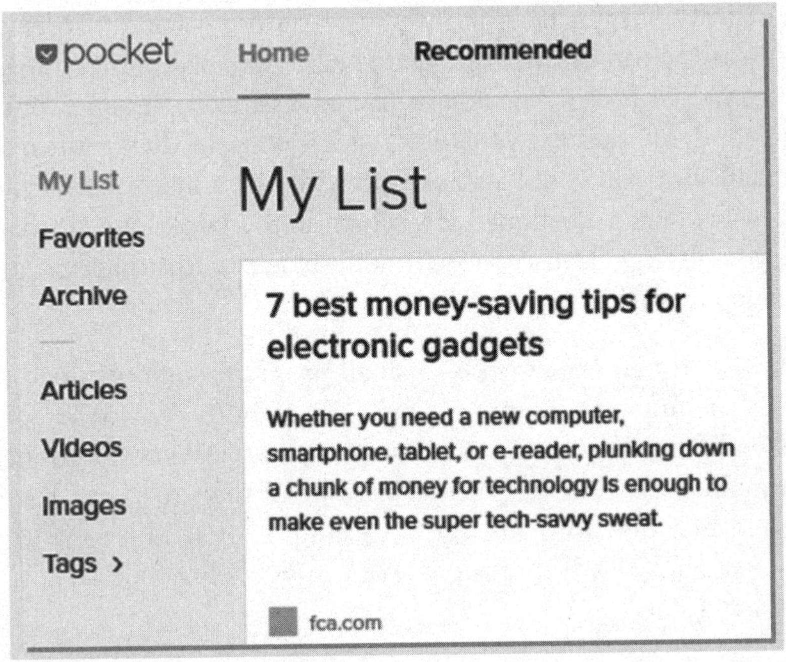

The best part is that most of the services allow you to read your saved articles on any supported device, so they're as portable as your favorite glossy magazine.

Print only what you want from the Web

Sometimes you just want to do things the old-fashioned way. Like print out an article you see online so you can browse through it later or share it with a friend who's not as computer savvy as you are.

But beware — printing a Web page can result in funky spacing and lots of images. Plus you'll use up a lot of colored ink on all those annoying ads. Thankfully, you have several ways to print the info you want without wasting expensive ink and paper on stuff you don't.

The easiest option is to look for a print button — or a printer icon — at the top of the page near the article headline. You might also see *printer friendly version* or *text only version*. An *email* option might send you the text you want, which you can then print from the email. But you could also end up with just a link to the article. Microsoft offers a Reading View (click on the book button) that gives you a fairly clean version of your article, depending on how the page is set up.

Otherwise, try an extension — a small program you download that extends the function of your browser. Readability (*readability.com*) and CleanPrint (*formatdynamics.com/cleanprint*) are good options. Printing cleanly from the Web is just one of their services. You can also save an article to read later, for example, or send it to your phone or tablet. Both extensions work in every browser except Microsoft Edge.

Quickly email a link to a friend

See a great article online you want to share? Most people copy and paste the URL into an email or look for an *email this article* button next to the Web article. But there's an even easier way to share interesting Web pages with family and friends.

Chrome. Press ⌘+Shift+i on a Mac or Ctrl+Shift+i on a PC. A new Mail message opens, and the body includes a link to the article at the top. The Subject line is filled in with the title of the article and the website. All you have to do is choose the recipient.

If you use Gmail in Chrome on either a PC or Mac, you can download the extension *Send From Gmail*. Once it's installed, a Gmail icon appears on your toolbar. Whenever you want to email an article, click the icon, type in the recipient's email address, and click *Send*. You can also send by clicking in the body of the email and pressing ⌘+Enter on a Mac or Ctrl+Enter on a PC.

Safari. You can add a Mail button to your toolbar. While in Safari, right-click (press Control and click) on the toolbar to choose *Customize toolbar*. Look for the envelope icon and drag it to your toolbar. Then click on it when you're in an article you want to share.

Edge. Click on the circular Share button in the upper right corner, then click on *Mail*. Choose your email account, type in the address, and click *Send*.

Easily find a single word on a Web page

Some Web pages are just text heavy. That's fine if you're really enjoying an essay on the history of air ducts. But if you just want a tidbit of information, scanning through massive chunks of text isn't necessary. Search by word or phrase within a specific Web page to get what you need.

No matter which browser you use, the keyword shortcut is ⌘+f for Macs and Ctrl+f on a PC. This opens a small search window, either at the top (in Safari, Chrome, and Edge) under the search bar or at the bottom (Firefox). Type the word or phrase and press Enter or Return. If the first instance of the word doesn't turn up what you want, pressing Enter or Return will take you to the next one, and so on.

If you prefer to use your mouse, go to the Toolbar and click on Edit > Find. In some browsers you'll have to click *Find* once more.

Speedy search shortcuts you need to know

Say you're looking at NASA's International Space Station website and want to learn more about astronaut Scott Kelly. You could do a search on NASA's site or look up "Scott Kelly" in your search engine. But why go through all that trouble? You can quickly dig up info on any word or phrase you see online with these easy shortcuts.

Chrome. Highlight the phrase "Scott Kelly," hold down the Option key, and drag it up to the Address bar. Or highlight it, right-click (or press Control and click), to view the drop-down menu that includes *Search Google for Scott Kelly*. (If Google isn't your default search engine, yours will show up here.) If you're using a Mac, you can also *Look Up* phrases or words to get a dictionary definition.

Firefox. Highlight the phrase "Scott Kelly," hold down the Option key, and drag it to the search box (the one with the magnifying glass). Firefox gives you the ability to search from here using any of your search engines. You can also highlight the phrase, right-click (or hold down Control and click), and use the drop-down menu.

Safari. Highlight the phrase, and while holding down Option, drag it up to the Address bar. Or highlight, press Control, and get the drop-down menu. You can choose *Search with Google* or the *Look Up* option (which may or may not turn up anything useful).

Microsoft Edge. Highlight the phrase "Scott Kelly," hold down the Alt key, and drag it up to the search box, or right-click and Windows' new built-in searching assistant, Cortana, will help. She'll pull up Wikipedia info about the subject and also offer to search Twitter, Facebook, or your default search engine.

Jump-start your search with keywords

Like to check facts on Wikipedia? Always looking up movie trivia on IMDb? Firefox has an awesome search feature that will save you time by taking you directly to your favorite search sites. All you need to do is create special keywords for those sites.

Go to the Wikipedia website — or almost any website with a search box — and right-click (Control and click on a Mac) in the site's search box. In the drop-down menu, you'll see *Add a Keyword for this Search…* In the dialog box that pops up, type in the keyword of your choice (a single word), such as "wiki" for Wikipedia or "IMDb" for Internet Movie Database.

Now whenever you want to do a quick search on Elvis, you can type in "wiki Elvis Presley" in the Firefox Address bar (known as the "awesome bar") and quickly find out when Elvis cut his first record.

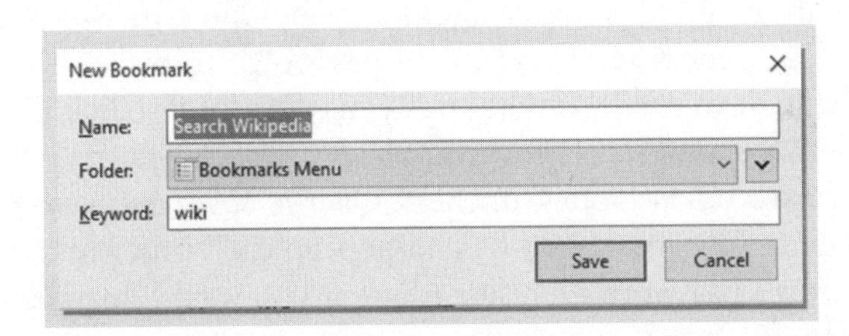

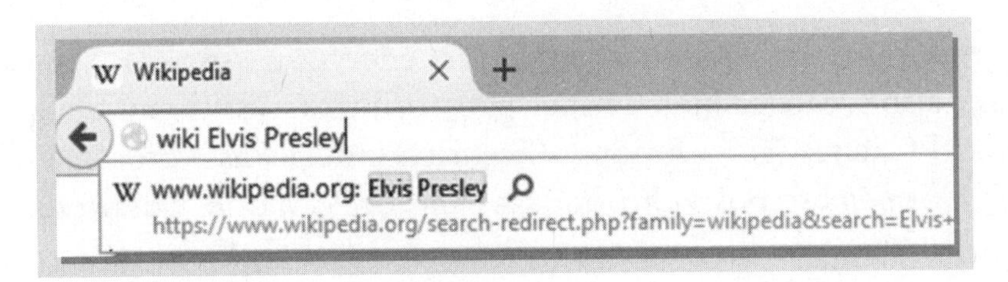

Let someone else remember your passwords

Trying to remember passwords for all your different websites is enough to give you a massive headache. But take heart. Here's an easy, safe, and secure way to keep track of them so you'll never forget a password again.

Trust your browser with password security. Many Web browsers offer to secure your passwords with varying levels of safety features.

- Microsoft Edge will save your website passwords by default. Just answer yes or no when it asks your preference when signing in to a website. If you don't want to use it, you can turn the setting off at ••• > Setting > View Advanced Setting > Privacy and Services. You'll see *Offer to save passwords* with a toggle button to switch on or off.

- Chrome gives you a save option, too. One criticism is that it doesn't encrypt, or code, your passwords to make them more difficult to crack. If that bugs you, turn it off. Click ≡ > Settings, and scroll down to *Show advanced settings.* Under *Passwords and forms,* uncheck *Offer to save your web passwords.* Otherwise you'll get a pop-up each time you create or enter a password on a site asking if you want Chrome to help you out.

- Firefox, on the other hand, does encrypt passwords. You can also create a single master password to access them. On a PC, click ≡ > Options > Security. Under *Passwords* check *Remember passwords for sites* and *Use a master password.* On a Mac, you'll click *Preferences* instead of *Options* and *Logins* instead of *Passwords.* You still have to create passwords for each website you visit, but from now on, Firefox will fill them in for you. Your master password is the only one you'll need to remember.

Firefox also features a free browser synchronization service called Sync. You can access your saved passwords as well as your Reading List, open tabs, history, and more settings, from any device running Firefox, including tablets and phones.

Add an app for stronger encryption. Separate browser add-ons and apps can manage your passwords with one master password, encrypt passwords, and generate stronger passwords. Some will even securely store personal information that you regularly use online, such as credit card numbers.

Blur (*abine.com*), Dashlane (*dashlane.com*), and LastPass (*lastpass.com*) are just a few of these password managers. Many have free versions in addition to annual subscription fees, with different options for each. Do some research to figure out which works best for you and will give you the most peace of mind.

Keep your private stuff private

You shared a hilarious family moment on Facebook, and now you're seeing comments on it by people you don't know. How could this have happened? Are you sharing private information online without knowing it?

Find out who sees what on social media. One free program, PrivacyFix, can show what you've been sharing and how much of it is public. Once you download the extension for Chrome or Firefox, a dashboard displays Facebook, Google, and LinkedIn settings that might be problematic.

For example, you may not realize you have Facebook set up to allow your friends to share your photos with their friends. PrivacyFix guides you through changing those settings to guard your private information. Download it from your app store or at *privacyfix.com*.

Get a browser extension to protect your info. Another extension available as a toolbar extension for Firefox or Chrome, Privowny (*privowny.com*) shows you which cookies and trackers are active on your favorite websites and allows you to block them. You can even use it to generate an anonymous email address when you register on a site to keep yours from being circulated and subjected to spam.

Use DuckDuckGo and surf anonymously. This alternative search engine could turn your browser — just about any browser — into the best Internet browser you've never heard of. It doesn't track, store, or share your browsing history, cookies, or any other personal data. Go to *duckduckgo.com* to learn how to make it your default search engine.

Wipe your info from the Internet. If you really want to clear your social, contact, and personal information from the Internet in a drastic way, software like DeleteMe periodically clears your publicly available information and photos from data sites — for a yearly subscription fee, of course.

Ban those pesky pop-up ads

Ads are a fact of life on the Web. But let's face it — they can be pretty obnoxious when they pop up and cover your Web page or blast you with irritating music and animations. And some of them even harbor malware, software that hacks into your computer and causes problems. Ready to fight back? Here is an easy way to get rid of those annoying pop-up ads — and it's free!

Install an ad-blocker plug-in for your browser. Adblock Plus is popular because it's been around for years and is available for almost all browsers. By default, it doesn't block what it terms "non-intrusive" advertising — ads that don't move, make noise, or interfere with your reading. But you can customize the plug-in to block just about any ad that you want.

To download Adblock Plus, go to *www.adblockplus.org*. It identifies which browser you're using and gives you a big Install button to click on.

Keep everything on your computer up to date. One way to fight back against harmful "malvertising" is to keep everything updated. This includes your browsers, browser plug-ins, and extensions. Updates often fix previously reported issues that can cause malware to creep in.

Stop advertisers from snooping

You're reading an article online when you notice an ad for a shirt you considered buying, but didn't. You switch to another website, and it shows up again. You feel like you're being stalked and, in a way, you are.

If an ad for a certain product follows you around online, you're the victim of advertising known as behavioral or personalized retargeting. At best, it's annoying. At worst, clicking one could lead you to download malware — software designed to harm your computer or gather sensitive personal information.

Companies track your surfing habits through the use of cookies, a bit of data that sticks to your browser. Cookies don't store personal info — and they're not always bad — but companies use them to figure out which ads best meet your interests. If you find these ads more creepy than helpful, here are some ways to get rid of them.

Block all Web trackers. Try an extension such as Ghostery, available for Chrome, Firefox, and Safari at *ghostery.com* or in your browser's app store. Ghostery might block things you do want to see, such as videos, but usually these extensions have adjustable settings to let you see only what you want.

Fight back against ads. Install the extension AdBlock Plus (see previous story). Other options include DoNotTrackMe, available for Chrome, Firefox, Safari, and Internet Explorer at *www.abine.com/donottrackme.html*; and Privacy Badger (*www.eff.org/privacybadger*, available for Chrome and Firefox).

Some ad blockers charge for their services and do more than just block ads, so it's worth researching to figure out which one works best for you. Keep in mind that some sites won't honor "do not track me" settings, regardless of what you use.

Clear out cookies and browsing history. This will keep nosy folks from knowing what sites you've visited on the web. But beware — cookies also are used to store passwords on your websites, so you'll have to go through the trouble of entering them again when you return to your favorite sites. With some browsers, like Chrome and Edge, you can pick and choose exactly what to delete.

- **Chrome.** History > Show Full History > Clear Browsing Data. Click on what you want to clear.
- **Safari.** History > Clear History
- **Firefox.** History > Clear Recent History
- **Edge.** ••• > Settings > Clear browsing data. Click *Choose what to clear* box, then check the items you want to delete.

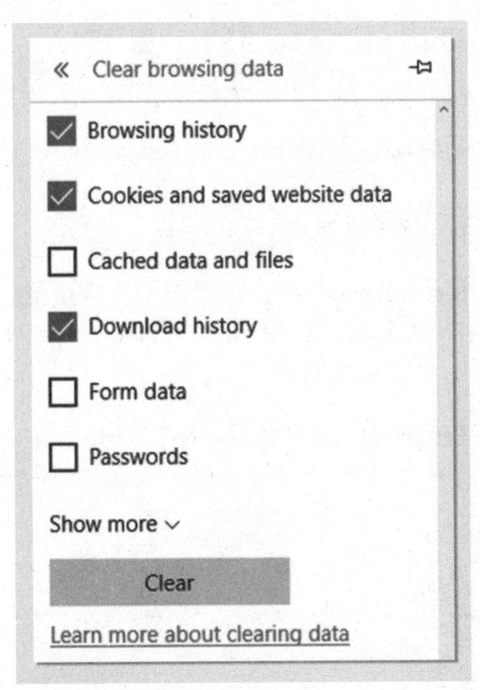

Go incognito while surfing the Internet

You find the perfect present for your spouse in an online store. The next day she goes on your shared computer, visits the same site, and sees the gift you'd bought. Oh no!

Usually it's helpful that your browser keeps track of where you've gone because it means your favorite sites load more quickly. But you could have avoided the surprise spoiler. One way is by clearing out your cookies and browser history. (To learn how, see the previous tip.)

Or you could be stealthy and use your browser's privacy, or incognito, mode. When you search this way, your browser won't store any cookies, passwords, images, search suggestions, or history. Just open up the privacy window or tab before you go buy that gift, close the window, and it's like you were never there.

Chrome. Open Menu (≡) > New Incognito Window. On a Mac, you can also press Shift+⌘+n. On a PC, Shift+Ctrl+n.

Firefox. Open Menu (≡) > New Private Window. Firefox lets you know that you're in Private Mode with a mask icon on the tab that reads *You're browsing privately.*

Safari. File > New Private Window. Or on a Mac, press Shift+⌘ +n. On a PC, Shift+Ctrl+n. In private mode, the search field will turn dark gray.

Edge. Open Menu (•••) > New InPrivate Window.

Google

What you don't know will surprise you

Free tech support at your fingertips

Has someone ever told you to "Google it" when you're trying to find information? Google has become synonymous with searching online because it claims more than 63 percent of the search engine market. And its Web browser, Chrome, has over 50 percent of the browser market. With stats like that, odds are you've used Google at some point in your life.

But have you ever used it to solve your tech problems? Next time your computer or device acts up, don't waste hours on hold with tech support. Try Googling your solution. It just might save you time, money, and hassle.

More than likely, you'll find that many people have had the same problem. Of course, knowing you're not alone doesn't help. That's why you use a special Google search hack. Just add "solved" to your search description. Google will only bring up results for questions that have been resolved.

Learn to speak Google and get better search results

You may not like it when your friends ignore the rules of punctuation. But don't be surprised when Google does it. Search engines ignore most punctuation and symbols. But some of those special characters might actually help you Google more effectively and get the search results you're looking for. Here are some to try.

Quotation marks. When you search for several terms at once, Google gives you results that contain the words anywhere on the Web page. The search terms could be in the same sentence or three paragraphs apart. If you need the exact phrase, put quotation marks around the words. Google will then search for them together and in that order.

Dash. You can exclude terms from searches with a dash (minus sign). This comes in handy when you want a recipe without certain ingredients. Or if you search for a word that has several different meanings.

For instance, you can't search "saw but not the movie" and get information on a new saw for your tool collection. Google focuses on the words "saw" and "movie" and only brings up those results. But if you type "saw -movie," Google will weed out the horror-film results.

Asterisk. If you're searching for a specific quote or phrase but can't remember all the words, you can type Google's wild-card symbol, the asterisk. It doesn't always guarantee better results — for example, "* if by land * if by sea" may provide the same answers as "if by land if by sea." But you may get more refined results for a query such as "All that * is not *."

Don't rummage around the Web — search specific sites

Domain. It sounds like a mythical realm or mighty kingdom. And when it comes to the Internet, it may sound like a foreign concept. But you are actually more familiar with domains than you think.

A domain is an Internet address, like *google.com*. You use domains all the time. But you probably never thought to use Google to search in them.

Search a type of site. You can search top-level domains, like .org, .gov, .edu, and .com, to get results from different types of sources. Just add "site:" before the top-level domain.

Suppose you want to find out how to stay youthful and vibrant. Of course you want information from credible sources rather than some health nut who has never read a scientific study. If you search "prevent aging site:.gov," you will only see results from websites that end in the domain .gov — including government sources such as the National Institute on Aging and the United States National Library of Medicine.

Search a specific site. Maybe you just want some tips you can use to help prevent aging. You can narrow your search even more by searching within a whole domain. For example, to search FC&A's website for aging tips, type "prevent aging site:fca.com." Google will search within *fca.com* for the terms. This is also handy if you want to search a website that doesn't have a search box.

Exclude a website. Sometimes you want to exclude a website from search results entirely. For instance, if you want to learn about aging prevention but don't want to see results from *Prevention* magazine's website, search "aging prevention -site:prevention.com."

Keep Google from following your every move

You may not realize it, but when you're logged in to your Google account, your searches and browsing activity are being saved in *Web & App Activity*. One reason Google keeps tabs on your searches is to improve suggestions and offer personalized content.

If you want a little more privacy, you can delete search activity one item at a time or all at once by signing in to your Google account and going to *history.google.com/history*.

Remove one by one. Check the box next to the item you want to delete. Or check the box next to the date, if you want to delete everything from that day. At the top of the screen, select *Delete*.

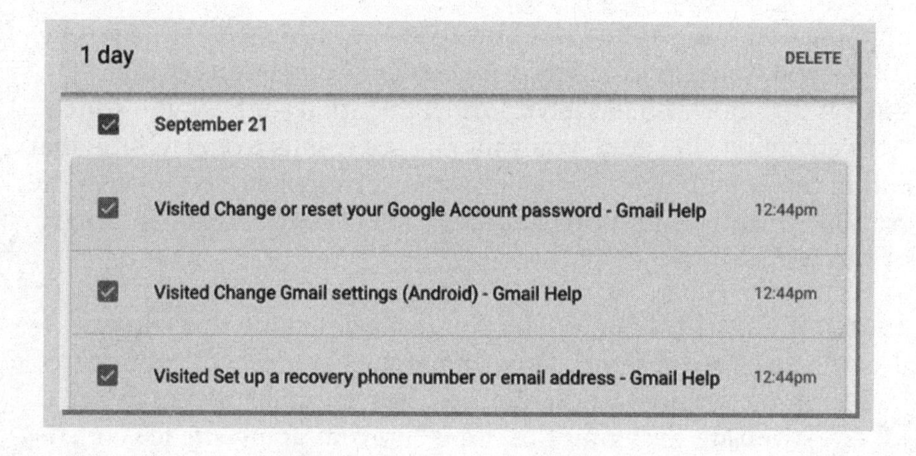

Erase all at once. Click the three vertical dots icon in the Menu bar as shown on the following page. Select *Delete options*. To delete everything, select *Advanced*. Under the *Select date* drop-down menu, select *All time*, then *Delete*.

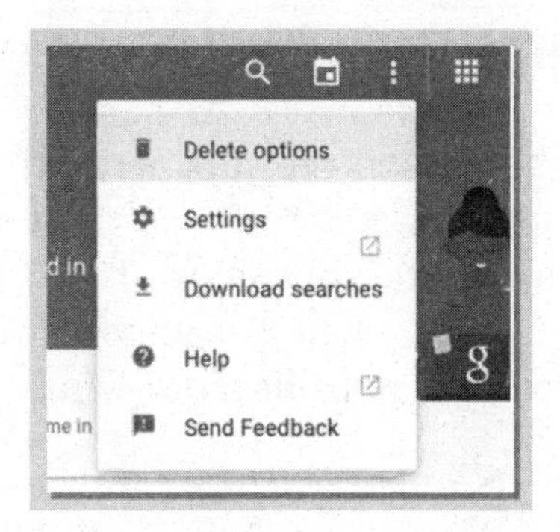

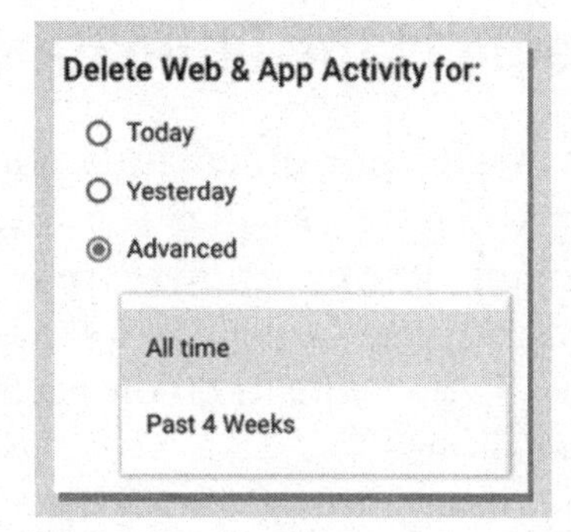

Stop future searches from being saved. You can pause *Web & App Activity* by switching it off at *google.com/settings/accounthistory/search*. This won't end the relationship though. Google may still store searches temporarily in order to improve the quality of your active search session.

Just a note — your Google history is not the same as your browser history, which you access through *History* in your browser toolbar. Find out how to clear recent history in your browser on page 237.

Filter down your results to the files you want

You're looking for a face-mask recipe in pdf format so you can save it for future reference. But when you do a search, you're swamped with pages of results that may not even be pdfs. Instead of wading through all those results, use this quick tip to get the file type you're looking for.

You can search for certain file extensions like .doc (document), .pdf (portable document format), .ppt (PowerPoint), and .xls (Excel spreadsheet). Enter "filetype:" followed by the three-letter file extension, like this — "face mask recipe filetype:pdf."

Google is the ultimate caller ID

Why waste your time with a phonebook (and risk a paper cut) when you have Google? It's always up-to-date. You won't lose it in your junk drawer. And you're not limited to your neck of the woods because Google is worldwide. Plus, when a strange number shows up on caller ID, you have an easy way to find out who's been calling.

Just type "phonebook:" into the Google search field followed by the mysterious phone number. Google will search through any information available online, showing you results that might include where the phone is registered, who it's registered to, and even which phone service your mystery caller uses. Bet your phonebook can't do that.

14 things you never knew Google could do

Many people think of Google as a place to answer all of your tough questions, like "why did I get married?" and "how do I make my cat love me?" But Google can do much more than that. Park yourself at *google.com* and try your hand at these cool features.

Define a word with the ultimate dictionary. In the search field, simply type "define:" followed by the word you want to look up. You'll see a box with the definition of the word, how to pronounce it, its origin, and more.

Set a timer for the final countdown. Just tell Google how much time you need. Your search will look something like "5 minute timer." If it doesn't kick off automatically, select *Start* to begin the visual countdown. An alarm sounds when time runs out. You can stop, reset, and even make it full-screen.

Translate languages you didn't even know existed. Search something like "translate amazing to spanish." You have options such as translate by voice, listen, and swap languages as well as a choice of 90 languages. Or you can go straight to *translate.google.com*. If you want to translate a whole Web page, paste in the web address, and click on the address it spits out. If you're on an Android phone, you don't even need the Internet to translate.

Compare food nutrition instantly. Pit your foods against each other by performing a search like "apple vs orange." Google shows you how healthy (or not) one food is versus another with a comparison of nutritional information including calories, sugars, and vitamins.

Solve math problems like a prodigy. Type in your equation. It's as simple as "84*3/6" (* means times, / is divided by). Google displays a calculator with the answer. Plus, now you can use the calculator — even for complex equations.

Calculate tips like you're always this organized. This is especially helpful when you're splitting the tip. Just type the bill amount like this: "tip $25.70." Don't forget the dollar sign. A tip calculator figures up the tip and allows you to customize according to the amount of the bill, tip percentage, and number of people.

Know a storm is coming before the first raindrop. Worried about rain for your weekend picnic? Search your local forecast, for example "weather savannah ga." You'll get results that show the place, day, month, temperature, and weather conditions as well as the forecast for the week.

Convert units in a matter of seconds. Plug in conversions like distance and temperature. Simply type "13 miles to feet," and you'll see the unit conversion.

Translate Roman numerals like an Italian. Easy peasy — just tell Google what you want. For example, "731 in roman." Google replies with a box that contains the numeric number translated into Roman numerals. Translating a Roman numeral into a number is not quite as quick, but Google will provide websites to help.

See what time it is anywhere in the world. Find out what time it is anywhere with Google's world clock feature. Just search for something like "time bora bora." The results will show the time and date of the place in question.

Convert currency during your world travels. Type the currency you want to convert, for instance, "58 dollars to euros." You'll get a box with the currency conversion as well as an option to adjust the amount and type of currency.

Monitor your flight for less hassle in your travel. Search the flight number, such as "delta 1278." Instantly see flight details including the progress of the plane. You'll know if it's on-time or if you have time for another cappuccino.

Be your own travel agent and journey like a local. Decide where you want to go. Google understands queries like "atlanta nashville flights." One result will be a table of flights listed by price. Want even more info? Search at *google.com/flights*.

Track your package so you're always in the know. Enter your tracking number and click *Search*. See exactly where your package is. Whether it's in route or hanging out at the post office, you'll know when to expect it.

Search for images — with images!

Whether you want to remind yourself how awesome people look doing "the sprinkler" or figure out what a selfie is, Google probably has an image for it. Just go to *images.google.com* (or *google.com* and click *Images*), type in what you want to see, and press enter. You can filter your results by size, color, website, usage rights, and more. And Google has more tricks up its sleeve you may never have noticed before.

Take the mini camera icon located in the search field. This little icon has a major perk — the reverse image search. It's exactly what it sounds like. Instead of searching with words, you search with pictures!

Click on the camera icon, and upload a photo from your computer by browsing for it or dragging it into the search box. Or search with an image URL.

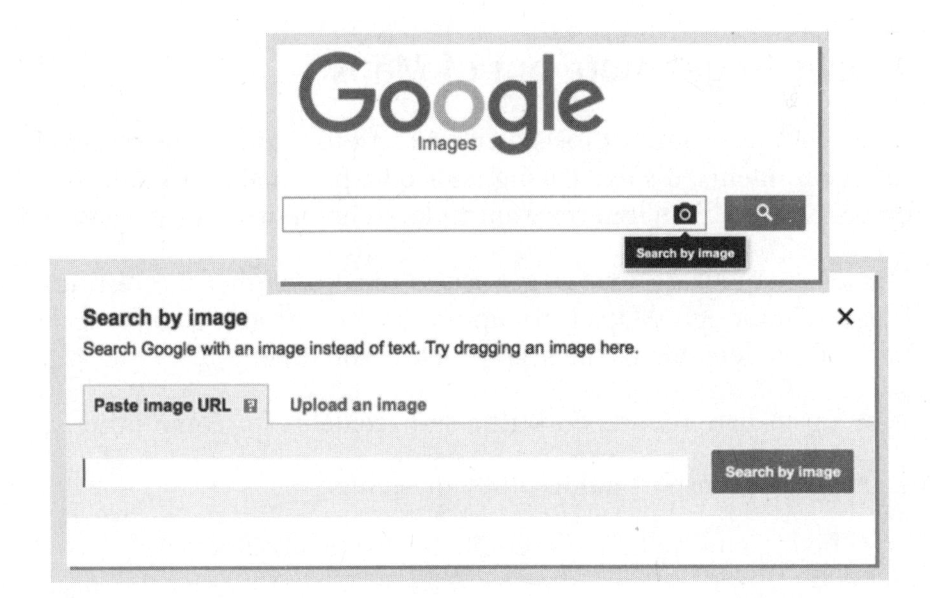

So now you may be thinking, that's interesting, but how is it going to help me? Here are a few scenarios to help you get started.

- You have a picture of your grandfather at a WWII memorial, but you can't remember where it is and what it's called. If it is a well-known memorial, there's a chance Google can identify it.
- You saved an image from an article you were reading online. Now you need the article, but you don't know which website it came from. Google will show you a list of Web pages with that image.
- You find a picture of some hiking boots you like on the Web, but it doesn't have any information. Google will show you similar images from all over the Web. One of them is bound to have the details.

4 ways to get more out of Maps

Google Maps is not just for driving directions. With imagery on all seven continents, it's like having your own personal tour guide — on your phone! You'll never want to leave home without it again.

One of the quickest ways to get access to all the other interesting features of Google Maps is to tap the menu button (three vertical lines on the left side of the search field). You'll get:

- up-to-date reports on traffic and accidents for commuting.
- public transport guides for sightseeing.
- biking and walking directions for being adventurous.

- search power for scoping out nearby restaurants and attractions. See a restaurant you like? You can check out ratings, reviews, price range, and even photos before making a reservation. If you have questions, simply call in with the *Call* button.

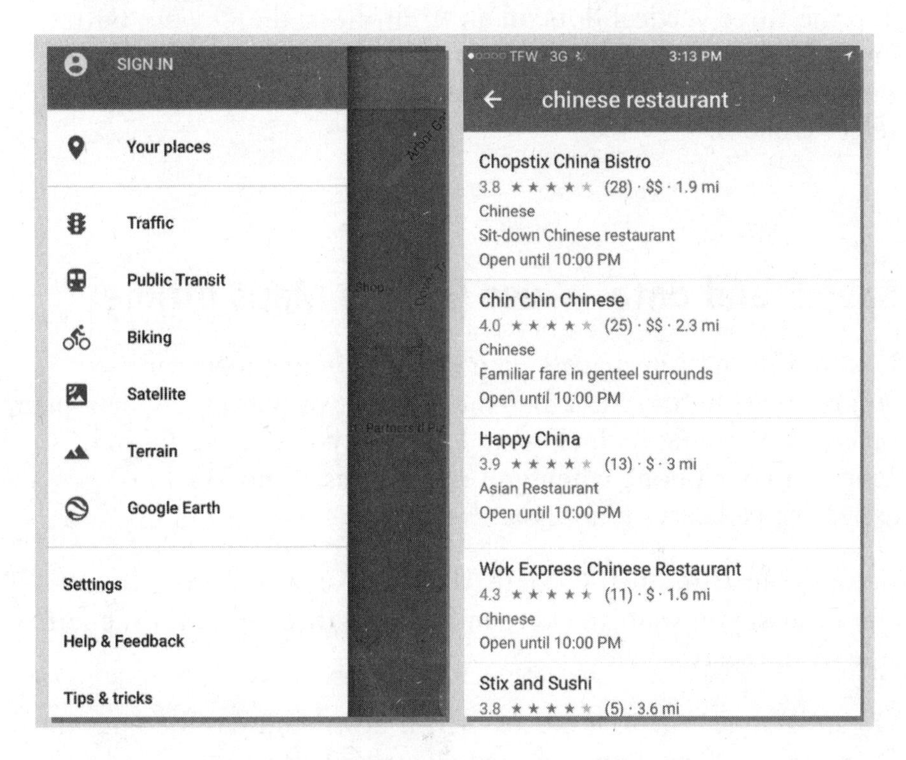

Simple way to steer clear of toll roads

When you're on the road, one of the last things you want is to be surprised by a toll booth. Google Maps has a feature just for people who don't like encountering the sudden toll road, ferry, or highway.

When you enter your starting point and destination, Maps automatically finds a route for you and lets you know if the route has tolls. This will look a little different depending on the device you have.

Tap the three vertical dots on an Android or the *Options* button on an iPhone. From the drop-down menu, tap *Route options* and *Avoid tolls*. If you're printing your directions from a desktop, select *Options*.

Save $ and data — use Google Maps offline

It would be great to have a map right on your phone when you're traveling overseas, but the thought of that pricey data plan makes you cringe. As luck would have it, you can use Google Maps on your phone when you're overseas — no Wi-Fi or expensive cell service necessary!

Before your trip, open Google Maps and sign in. Zoom in to the city or area you want to save, and tap the three lines on the left side of the search field.

On an iPhone, tap *Your places*. Scroll down and tap the option to *Save a new offline map*, then tap *Download*. On an Android phone, tap *Offline areas* then the + sign. On both devices, you will be prompted to zoom in if your map is too big. Name your offline map and tap *Save*.

You will find the map in *Your places* under *Offline maps*, or in *Offline areas*. Pay attention to when the map expires — you may have to update it within 30 days.

Ready to explore? Make sure you turn on airplane mode before you hit the streets. The new version of Maps on Android includes the ability to search and get directions, all without breaking the bank on data charges. Check your iPhone to see if those services are available yet.

When you want to update, rename, or delete the offline map, tap *View all and manage* on the iPhone, then tap the overflow menu (three vertical dots) next to your map name, and select your choice. On an Android phone, tap the map name, then the desired button.

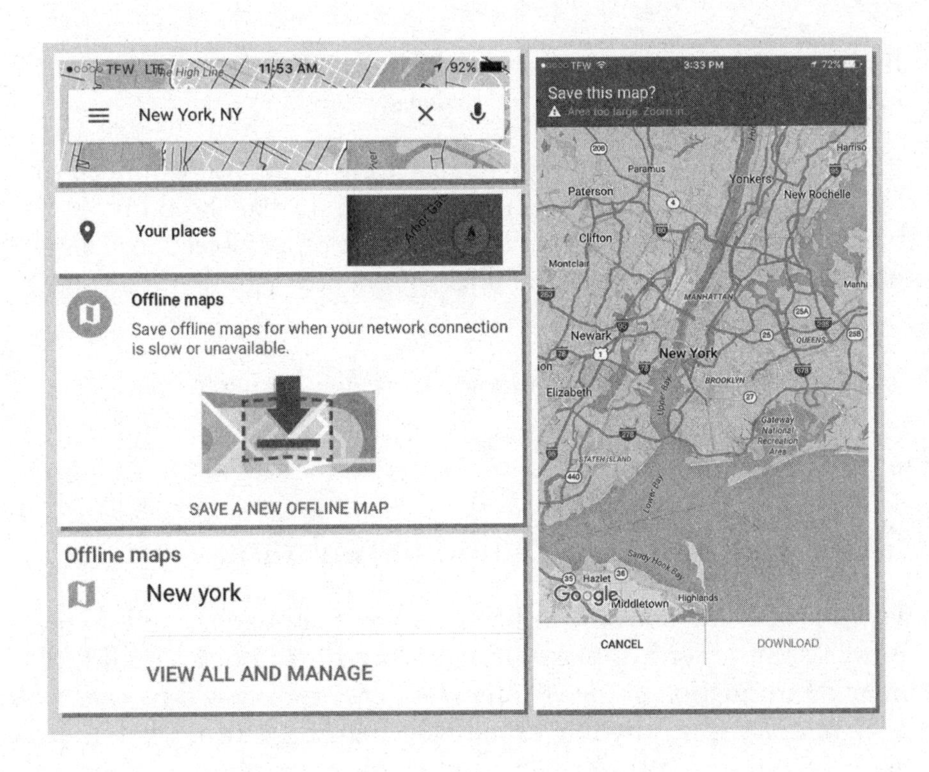

Hide your home from Google Street View

You may not have your own TV show, complete with paparazzi hounding your house, but you still want your privacy. If you don't want strangers to see your home from Google Street View, here's what you do.

On Google Maps, input your address and locate your home in Street View mode. You can do this by dragging the little yellow person on the bottom-right to the street in front of your house. (Don't worry, it won't hurt him.) If you can't drop him on your street, that means your house is not visible in Street View, so you're safe from prying eyes.

But if you see an image you want to report, click *Report a problem* located at bottom-right of your browser.

Fill out the *Report Inappropriate Street View* form. You will need to tell Google what's wrong with the image and what should be blurred — home, face, car, license plate, or other object. You also have to provide your email (and let them know you're not a robot) before clicking *Submit.*

That's it. Now it's all up to Google to review your request.

Share and edit files online in real time

Word processing and spreadsheet programs are handy. And if you need a comprehensive program like Microsoft Office, you'll find helpful information in the chapter *Word & Excel: expert ways to turn work into play*. But why buy expensive Microsoft Office when there are powerful word processing and spreadsheet programs you can use for free? Save your time and money, and check out Google Drive.

One of the major benefits of Drive is your ability to share files. With most programs, you have to email your files or save them in the cloud before you can share them with anyone else. But with Google

Drive, you can share your work online and collaborate in real time. That means you and others can all edit files at the same time.

From Docs to Sheets to Slides, you can see where your collaborators are within the file and what they're changing. And when you need to tell your coworker not to change your amazing paragraph on page 3, you can chat or leave comments within the doc. It's always up-to-date. And if you leave the file for a while, Google will show you what changed since you last viewed it.

The easiest way to share a file is to sign in to a Drive app, Google Docs for instance, and open the file you want to share. Click the blue *Share* button in the top-right corner. To see all your share settings, click *Advanced.*

You control who gets to view, edit, or comment on your file. You can share a link or send an email invitation by adding people by name or email address. The people you are working with don't even need a Google account.

Publish your files for all to see

The Google Drive sharing feature is great, but one thing it doesn't allow is sharing with a larger audience. If you want to share with more than a handful of people at once, publish your file to the Web.

Most of Drive's apps work in much the same way. From within your file, click *File* and select *Publish to the web* from the drop-down menu. Your file won't be published to the Web until you click *Publish*. After that, anyone with the link can view it as a typical Web page. Keep in mind, they won't be able to edit and comment.

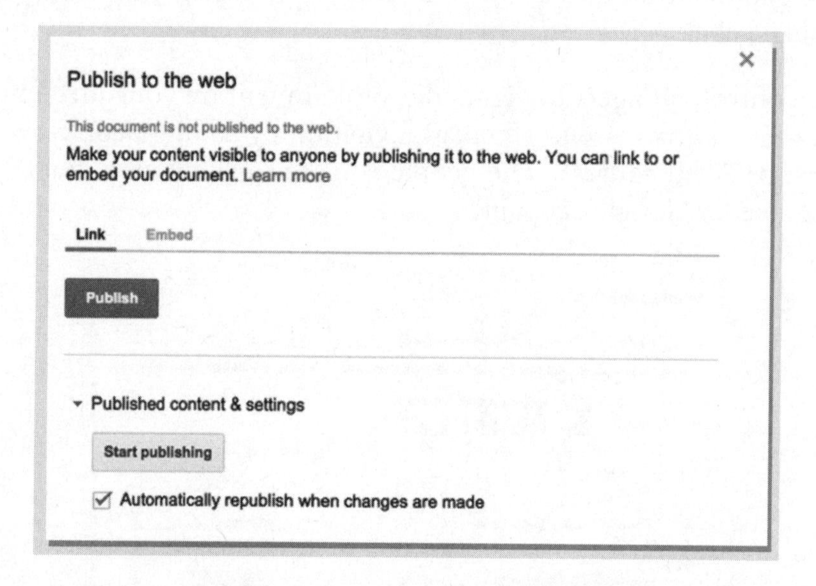

Updates to the doc republish automatically. To change this, go to File > Publish to the web > Published content & settings. Uncheck *Automatically republish when changes are made.*

To make the file private again, click File > Publish to the web > Stop publishing.

Navigate Google Drive like a pro

Google seems to do everything else, so why doesn't it have a storage feature like iCloud or Dropbox? Glad you asked. It does — and it's fondly known as Google Drive.

With Drive, you can store any type of file as well as access Drive apps such as Docs, Sheets, Slides, Forms, and Drawings. Here are three hints to help you get started.

Find Drive from anywhere. Finding Drive doesn't have to be an Easter egg hunt. Just open Chrome and sign in to your Google account. From the Google home page, you can open Drive with the Apps Launcher (9-square box in the top-right of your browser). Or get to Drive from anywhere by typing *drive.google.com/drive/my-drive* into the Address bar.

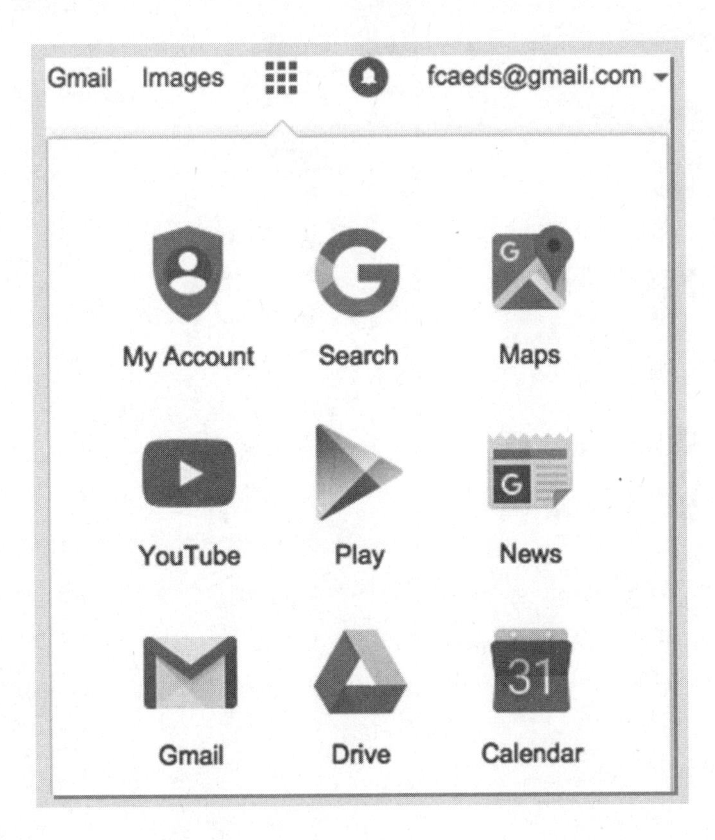

To return to Google Drive from a Drive app, such as Docs, click on the app menu (three lines on top-left) and select *Drive.*

Create new files right from Drive. To open a blank file with one of Drive's apps, click *New* and select the app you want to create a new file in.

Upload a file from your computer. If you already have a file on your computer, just upload it right into Google Drive or any of the apps. From Drive, click *New* and either *File upload* or *Folder upload.* Select a file or folder from your computer, or drag and drop, then click *Open.* If you're already in an app — Forms, for example — click on the folder icon in the top-right corner.

To learn about other cloud services that are available, see page 261.

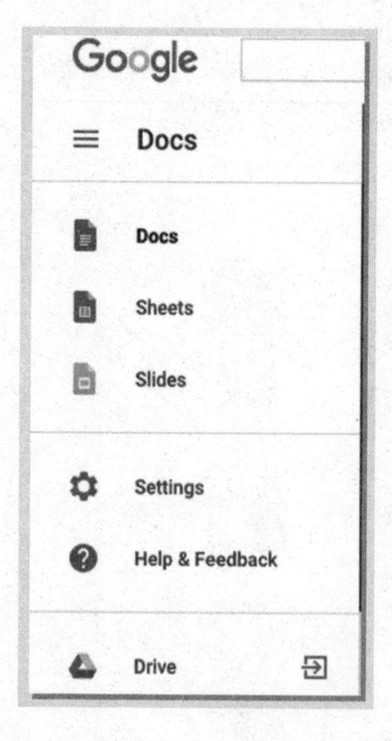

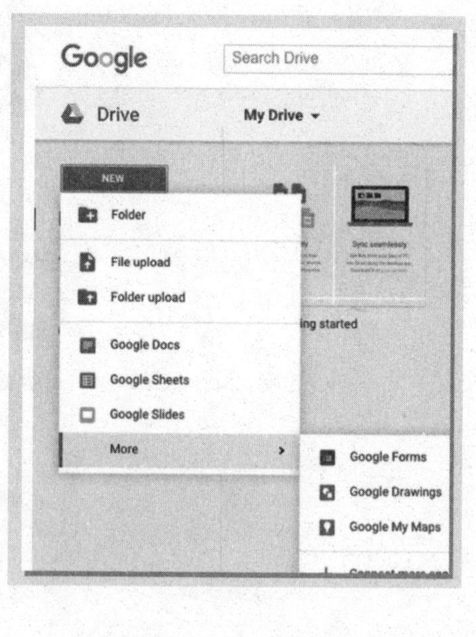

Work online files even when you're offline

Sometimes, you just don't have access to the Internet. Whatever the reason — tech glitch, power outage, zombie apocalypse — Google Drive has you covered. You can get to your files offline so you don't miss a beat.

Before you set off on your offline adventure, your computer needs two things — Google Chrome (*google.com/chrome/browser/desktop*) and the Google Drive app (*google.com/drive/download*), both of which are free to download.

Your Google settings should default to let you work offline, but if you have problems, make sure your settings are as follows.

- In Google Drive, click on the gear icon in the top-right corner. Select *Settings* from the drop-down menu. In the *General* tab, make sure *Sync Google Docs, Sheets, Slides & Drawings files to this computer so that you can edit offline* is checked.
- On the home screen of Docs, Sheets, or whichever apps you wish to use offline, click the Menu bar (three lines icon), then *Settings*, and make sure *Offline sync* is turned on.
- Once you're offline, you can open a file in your Chrome browser by selecting the file from the Google Drive app on your computer. Or you can open your Chrome browser and go to Google Drive at *drive.google.com/drive/my-drive.* Changes will be saved automatically, so next time you open Docs online, your offline changes will be synchronized. Crisis averted.

Search the Web without leaving your document

When you're working in a document, it's a hassle to keep flipping back to your browser to search for more information. Google Docs has the perfect solution.

Go to *Tools* and select *Research.* A modified version of Google search will open in a Sidebar, allowing you to search for information, images, quotes, and more — right from your document.

But that's not all. When you find something you want to use, you can preview the page before you open it, and you can insert images from the Sidebar by dragging them into your document. Plus, if you need to cite the source or insert a link, this tool will do it for you. Simply place your cursor over the search result you want to open to see those options.

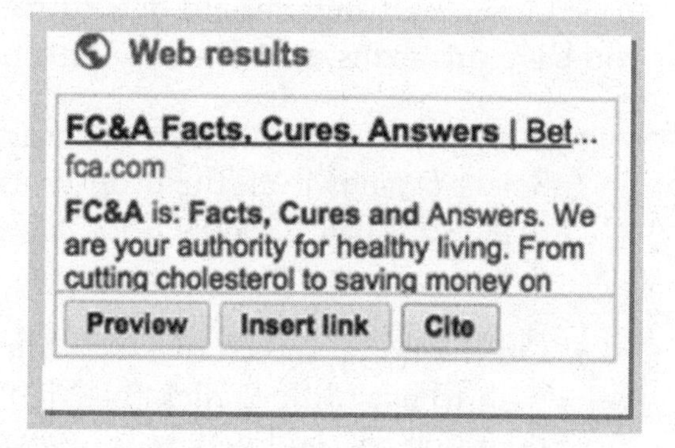

Google takes copy and paste to a new level

If copy and paste is already part of your daily routine, then you're going to love this. Google Drive has something even more magical that works in some Drive apps like Docs. It's called the Web clipboard, and it takes copy and paste to new heights.

The Web clipboard can hold more than one selection at a time. It hangs on to the selection for 30 days after you use it. And it works across computers — just sign in to your Google account.

Highlight or select what you want to copy. Go to Edit > Web clipboard > Copy selection to web clipboard. When you're ready to paste, make sure your cursor is where you want the selection to appear, and return to Edit > Web Clipboard.

Select what you want to paste from the list of recently copied selections. You'll see a preview of it below the box. Depending on the selection, you may be given different paste options, such as HTML or plain text.

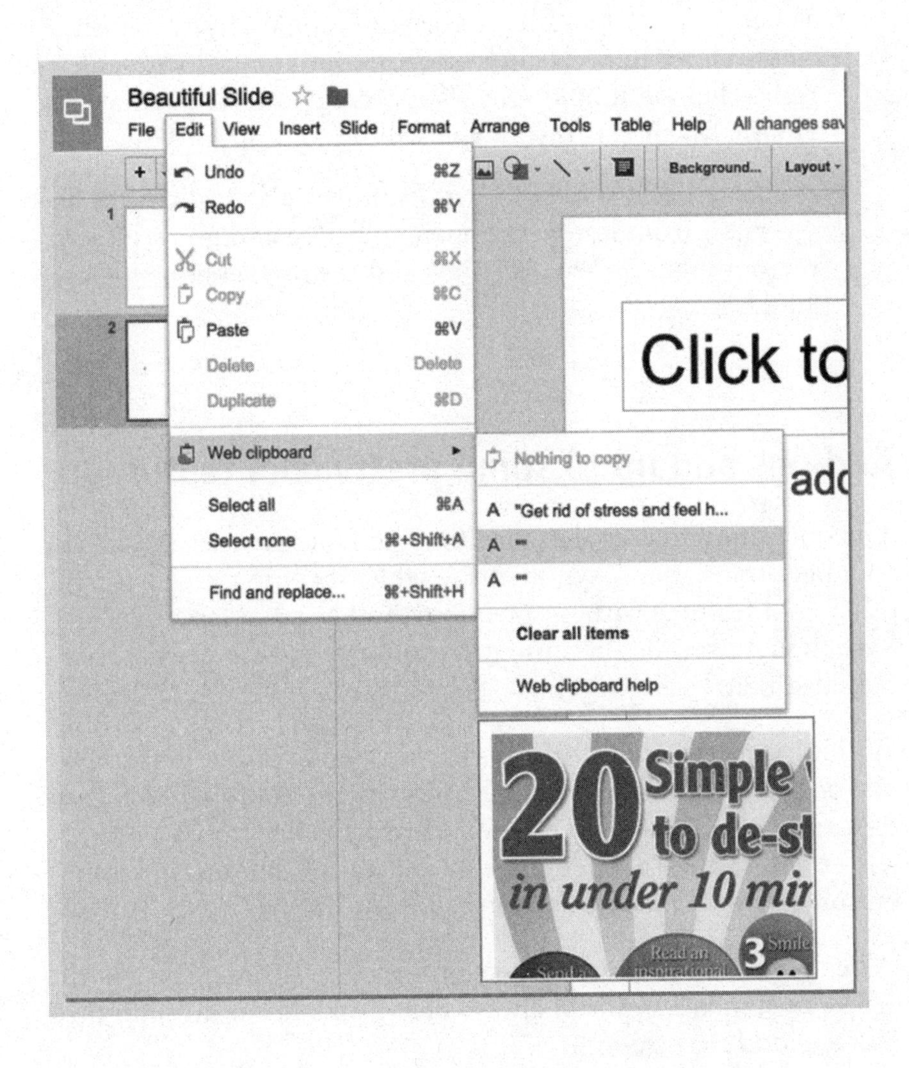

2 ways to recover files from Drive

Oops. It happened again. You need that file. The one you deleted last week. But now what? Is your document lost forever?

Not quite. Here are two ways to recover a document when it seems all is lost.

- When you delete a file, it goes to Google Drive's Trash. But all you have to do is click on *Trash*, find your document, right-click on it, and select *Restore*. As long as you haven't emptied your trash, your file will be waiting for you.
- If you've been working on a document and realize you need a version from last week, open your doc and go to File > See revision history. You can view and restore versions for up to 30 days.

Add-ons add up to some pretty cool features

The only thing that could make Google Docs any better is if you could get translations, synonyms, a table of contents, charts, and other cool features without opening another tab. Google must have thought so, too, because it created the add-on feature that does just that.

Within your doc, go to *Add-ons* in the Menu bar and select *Get add-ons*. You will have the choice of many useful tools such as a thesaurus, translator, and citation creator. When you find one you're interested in, click on it to get more info, or click the button *+ Free* to place it on your add-ons list.

When you want to use your add-ons, they'll be right there in the drop-down menu. You can always delete an add-on through the *Manage add-ons* option.

Apps

Online services that can change your life

Pick the cloud service that's right for you

If your head is in the clouds over "the cloud," then it's time to clear up the confusion.

The cloud refers to a network of servers on the Internet where you can store and manage files and photos. Cloud storage companies use these remote servers to save your data, which you can then access from any device with an Internet connection.

Why would you want to use the cloud? Perhaps the most important reason is to safeguard your important documents, photos, and music collection. If your computer, smartphone, or tablet crashes or gets lost, you can access your information from another device as long as it's connected to the Internet.

But wait, you say. I have an external hard drive and flash drive that do a great job of backing up my computer and pictures. So I'm covered, right? Amy Rivers thought that's all she needed, too, until one day her portable hard drive was stolen out of her car.

She lamented on Facebook that she had saved years of family photos on that hard drive, only to lose them all. If she had uploaded them to the cloud — her virtual vault in the sky — she would still have them today.

Worry-free storage is just one of the many benefits the cloud offers. Here's a quick look at the most popular cloud services with a few of their amazing features. Most offer free, limited storage space with more space available for a fee. Simply go to the cloud service's website, register for an account, and go through the installation steps.

iCloud. Don't let the name of Apple's storage space confuse you. There's "the cloud" which is the Internet's storage space. Then there's Apple's iCloud, one of the many cloud services available to you.

Think of iCloud as an apple tree and all of Apple's apps as the fruit dangling from its branches. For instance, the iCloud Photo Library app gathers your photos and videos, and iCloud Drive stores your documents. Apple's Family Sharing let's you share your music, movies, and photos with up to six family members. (See page 70 for more about Family Sharing.) And if you lose your iPhone, the iCloud's Find My Phone feature makes it easy to track it down.

Plus, iCloud synchronizes calendars, messages, and reminders from all of your Apple devices. And if you don't own a Mac computer, no worries. You can download the iCloud app for your PC. Best of all, 5 gigabytes of storage space are free. How 'bout them apples!

OneDrive. OneDrive is Microsoft's version of iCloud. It started out as a way for Android and Windows users to sync and store files created in Word, Excel, OneNote, PowerPoint, and Adobe Photoshop and Illustrator.

OneDrive now supports Microsoft's Office Online suite. So you can access your files from any computer, share them with friends or coworkers, and invite multiple people to add to or edit your files. Plus, OneDrive's got an app for Mac and iOS. Store up to 15 gigabytes free. Or pay a few bucks a month for extra storage.

Dropbox. Share hundreds of photos with anyone you want to — without having to email them — by saying hello to Dropbox.

Dropbox loads easily to any device and is used by millions of people worldwide. So there's a good chance the people you want to share with already have an account. And it's free, too. Dropbox offers 2 gigabytes of storage free, but you can earn more by referring friends.

Google Drive. When it comes to storage, backup, and sharing, Google's got it all. Google Drive makes it easy to save Gmail attachments, create photo albums, and share Google documents, spreadsheets, and presentations. Plus, you can save Microsoft Office files, too.

And with the Drive app for Android, you can snap photos of any document like a receipt, and Google will store it for you as a PDF. Google Drive is also available as an app for any smartphone, tablet, and computer. You get 15 gigabytes of free storage, or you can pay for additional storage.

Amazon Cloud Drive. You've probably ordered a book from Amazon and maybe even CDs, DVDs, electronics, and a whole host of other merchandise. After all, it's the world's largest Internet retailer.

If that weren't enough, Amazon now offers Amazon Cloud Drive for storage and backup in two plans — Unlimited Photos for photos only, and Unlimited Everything for both photos and documents. Unlimited Photos is free for Amazon Prime members. Otherwise, Amazon Cloud charges an annual fee.

Easy way to share Dropbox files

Dropbox makes it easy for you to share photos with a friend who doesn't have an account.

Sign into your account at *dropbox.com.* Put your cursor on the folder you want to share, and you'll see a *Share* button on the right. Click on it, then choose *Send link*. Dropbox will create a Web link for your folder.

Type your friend's email address in the space provided and click *Send*. Your friend will get an email with the link, which she can open from her Web browser. Doesn't get any easier than that.

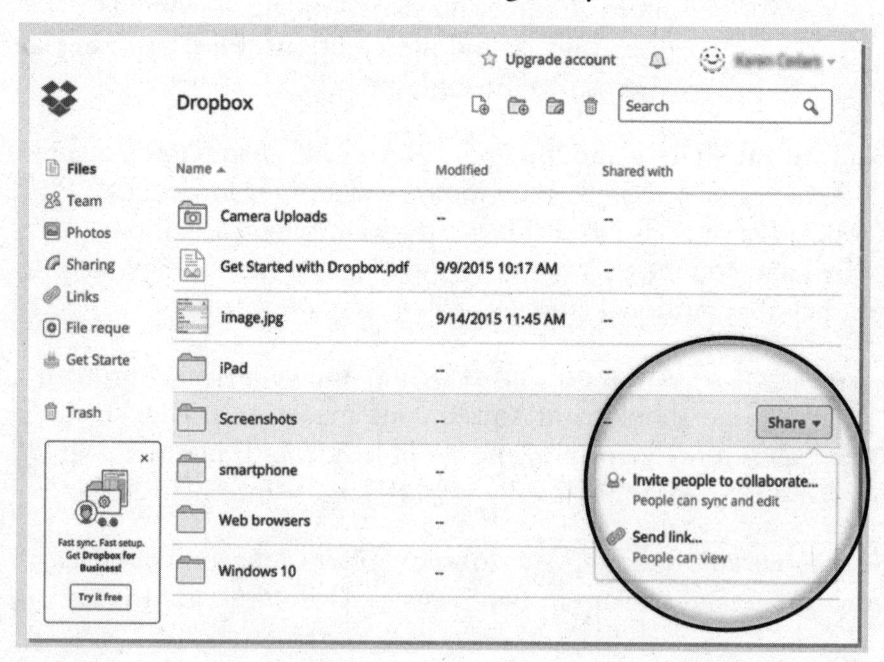

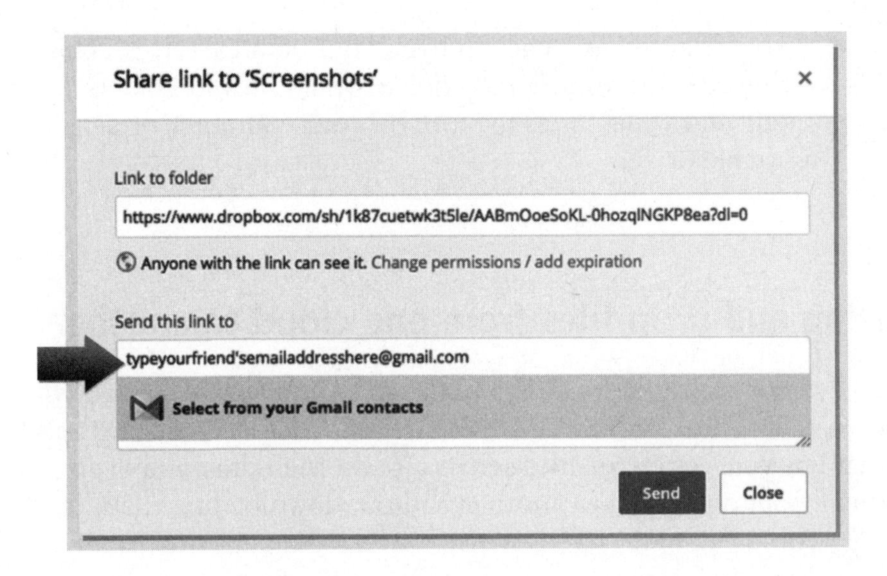

3 ways to protect your cloud files

Should you store your documents in the cloud? Yes, the cloud is a great place to store anything from documents to photos. But it's not bullet proof. Like most anything online these days, hackers can get into your files and steal sensitive information. Here are three ways you can protect yourself.

- Add an extra layer of security to your cloud storage with Boxcryptor at *boxcryptor.com*. The program allows you to encrypt any file you store on major cloud storage providers like Dropbox, OneDrive, and Google Drive. And it works on Macs, Windows, Android, and Apple devices. The basic program is free for personal use, or you can pay an annual fee to download Boxcryptor on multiple devices and get online support.

- Transfer your most sensitive files to a flash drive, and delete them from your computer and cloud account. Make sure you store the flash drive in a safe place.

- Turn off automatic backup from your cloud services. This will force you to pick and choose which files to save to the cloud and which ones to store on your computer or a portable drive.

Drag and drop files from one cloud to another

You've got files in iCloud, Dropbox, and OneDrive. How do you keep them all organized? With *Otixo.com* — a file-managing app that lets you search for files across clouds, then drag and drop them from one cloud to another without downloading them to your computer first.

Take a look at the example. The Dropbox file titled Snap.png is being dragged and dropped into the Amazon Pictures folder. It's that simple.

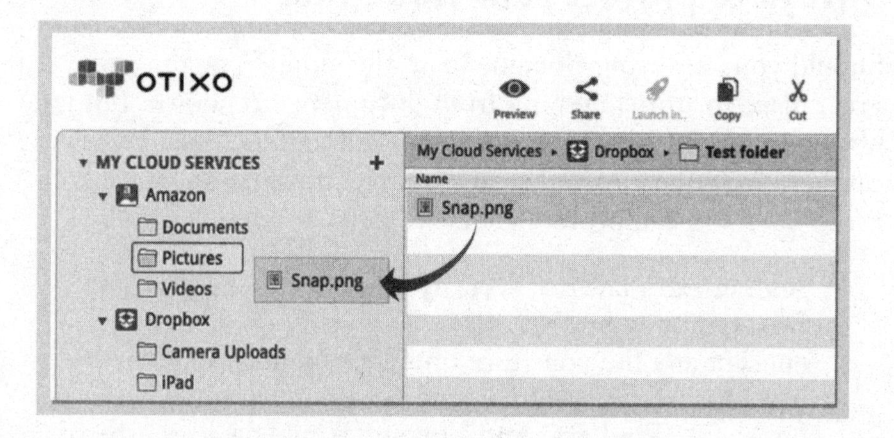

Otixo does not permanently store your files. It's a tool to access all your files across multiple clouds in one place. The program is free for basic service, or you can pay a monthly fee for additional features. Both the free and fee-based services include unlimited file transfers. Otixo also offers business accounts.

Easy way to find hidden apps

Shopping for apps is as easy as looking for candy canes at Christmas or chocolate bunnies at Easter. If you're an iPhone or iPad user, look for the Apple app store icon on your device.

And if you've got an Android phone or tablet, look for the Google Play store symbol.

Once you've opened the store, use the search function to look within an app category, for example, budgeting or fitness; or to find a specific app by name, like Scrabble or Gas Buddy.

Occasionally, you may not find the app you're looking for, but you know it exists. Most apps have a corresponding website, so open your device's Web browser and search for the app's website. Then look for the Apple or Google Play links.

Available on the App Store   GET IT ON Google play

Tap the one that goes with your device and click *Get.* Once it's installed, it's time to play. Have fun!

Pay less for your prescriptions

Stop paying an arm and a leg for your medications. Get the lowest price on your prescriptions with a little help from the Internet. Digital discount programs such as LowestMed let you compare drug prices in your area then download a discount card to take to your local pharmacy.

Visit the website *LowestMed.com* or download the LowestMed app on your smartphone. Click or tap the *Search* button, enter a drug name, and select your medicine from the list that appears. The website will also ask for your ZIP code. Almost immediately, you'll get a list of drug prices at pharmacies near you. Check the dosage and change it if needed. Find a price that's right?

- On your smartphone, tap *Show Card* next to a pharmacy's name, and a discount card will appear. Show this to the pharmacist when you pick up your prescription. Otherwise, you won't score the low price you saw in the search results.
- On your computer, click *Show Card* beside the pharmacy you pick. Click *Print the card* on the next page, and take that card to the pharmacist.

Free medical advice is just a click away

If your doctor doesn't make house calls — and who does these days? — you can try the next best thing. Ask a real doctor a question online for free. Check out *HealthTap.com*. You'll get an answer from one or more doctors in about 24 to 48 hours. While you wait for your answer, HealthTap will show you similar questions with actual physician responses.

And if you want a more personalized experience with the doctor of your choice — you have a whopping 73,000 U.S. doctors to choose from — you can pay a fee for a video chat or private message exchange. Wait time is about 24 hours.

Need immediate attention? Pay for the prime service, and you'll get a speedy response to your questions from a doctor on call. Response time is usually two minutes. Think of HealthTap as your very own digital doctor.

2 apps make end-of-life planning easier

You know you don't want to be kept on life support if something happens, but you need a way to share your wishes with your loved ones. Two services make the process painless — the My Health Care Wishes app and the Everplans website.

- My Health Care Wishes was developed by the American Bar Association. It lets you store your end-of-life directives directly on your smartphone and share them with medical personnel via text, fax, or email. You can also sync the data to your Dropbox account and store it in the cloud. You can update your directives at any time as you get older.
- Everplans helps you create your directives online by answering a series of questions. Then you designate deputies — people you trust — and share details of your Everplan with them. Get started at *Everplans.com*.

Whichever service you choose, you'll rest easy knowing your wishes are just a couple of clicks away.

Find top-notch doctors with online reviews

Looking for a doctor is like looking through a box of chocolates — you never know what you're going to get. Thanks to three websites, though, you can arm yourself with patient reviews to find the best one.

Check out *Vitals.com*, *Healthgrades.com*, or *RateMDs.com*. Enter a doctor's name, medical specialty, or a health condition along with your location. You'll find reviews on doctors and office staffs, the services and procedures they offer, and which types of insurance they accept.

No more chasing down your medical records

Are your medical records scattered across half a dozen doctor's offices? They don't have to be if you use *HealthVault.com*. It's a great way to have access to your medical records any time you need them — even share them with a new doctor without the hassle of numerous phone calls. And it's a lifesaver to boot.

HealthVault lets you gather all your health information in one place online. So when you hop from one doctor's office to another, you take your records with you via your cellphone or tablet. What's more, HealthVault offers a wide range of apps with an amazing array of capabilities. For instance, one app lets you store and share X-rays and CT scans. Plus, you can set up hundreds of devices — from blood pressure readers to blood glucose monitors — to load data directly into HealthVault.

And there's one other feature that could save your life — the emergency profile. It shows first responders your most important health information. Best of all, HealthVault is free.

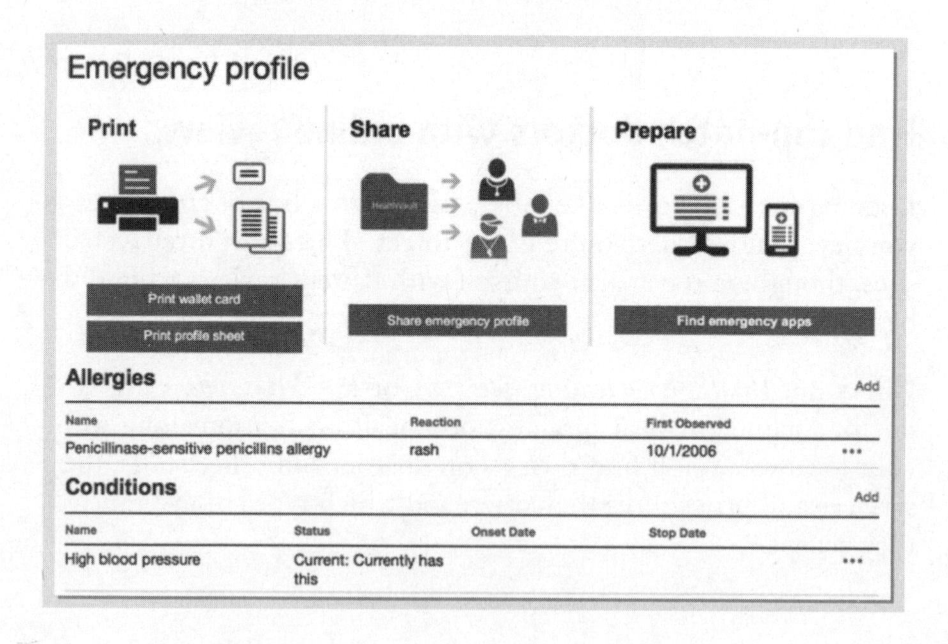

App makes losing weight a piece of cake

Let MyFitnessPal help you shed a few pounds. By signing up for an account at *MyFitnessPal.com* or loading the app on your smartphone or tablet, you can keep track of your weight loss and fitness goals, get nutritional info, find recipes and health blogs, and enter reminders. It also lets you connect with friends for moral support.

And make sure you use the app's barcode scanner with your phone or tablet. It's a great way to keep track of portion sizes. To use it, pick *Food Diary* then choose *Add Food* under Breakfast, Lunch, Dinner, or Snacks. Next, look for the barcode symbol on the top, right-hand corner of your device. Tap it and pick *yes* when it asks if it can access your camera.

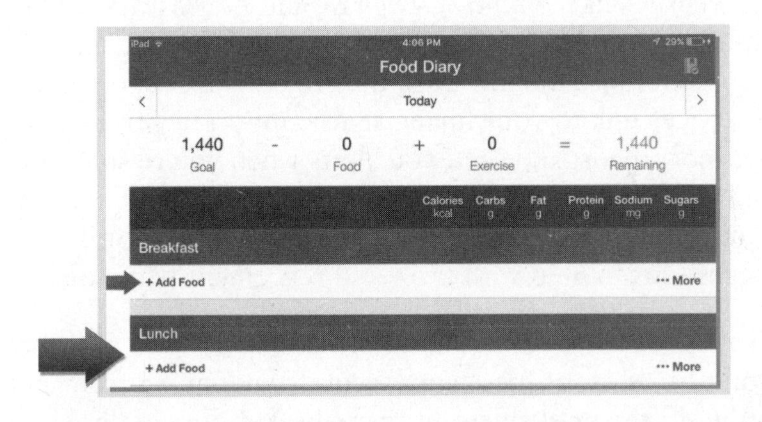

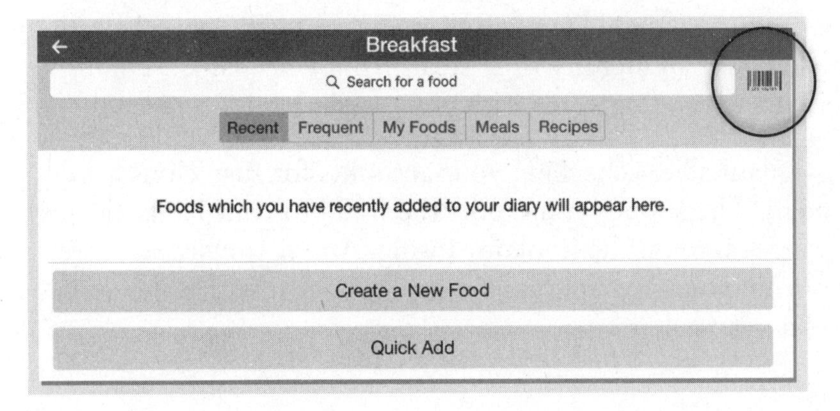

Hold your device over the barcode of any packaged food, centering it within the square that appears on your screen. The app will load the food's nutritional data, including calories. Just enter how many servings you had, and tap the check mark to add it to your food diary.

Apps to make budgeting fun — really

Ah, the joys of budgeting. Don't you just love keeping track of every single penny you spend? Of course you don't! That's why you should let an app do it for you.

Apps and websites like *Mint.com*, *Digit.co*, and *LevelMoney.com* offer simple and low-cost ways to keep track of your spending, saving you even more money. Plus, they're easy to set up. Some of these services link to your financial accounts. They track your spending, spot trends, and send you alerts when you're splurging. Others move a little money from your checking account and stash it for you in a savings account before you can spend it. Still others make suggestions to help with taxes and retirement planning.

Worried about sharing your bank information with one of these sites? Don't be. Most of these apps securely encrypt your financial information. In the unlikely event that a hacker does get hold of your money, federal laws will protect you — the same laws that shield you if an identity thief steals from your bank account or goes on a spending spree with your credit card.

Be vigilant about checking your accounts for suspicious activity, though. The sooner you report the fraud to your bank, the less money you're on the hook for losing. And if you access these services from your smartphone, be sure to lock it with a secure code in case it gets stolen.

You can also check out services that don't link to your bank account, such as *YouNeedABudget.com* (YNAB). Unlike the other apps, you enter all your bills and receipts into YNAB manually. YNAB is free for the first 34 days. After that, you'll have to pay a one-time fee to download the software. Mint, Digit, and Level Money, on the other hand, do most of the work for you and are free.

Get the most out of Mint's features

Be sure to sync all of your financial data, including car loans and real estate assets, when using the Mint budgeting app. In return, Mint will assess your budget and net worth more accurately. And use the goals feature for paying down loans or saving for big purchases. It will make suggestions based on your objectives. For instance, if you want to buy a new computer, this feature will tell you how much money to set aside each month to reach your goal.

Make filing your taxes a breeze

Christmas may be the most wonderful time of year, but what comes after is the most painful — tax season. Ouch! But not anymore. A tax preparation service is about to make filing your taxes a breeze. It's called TaxAct Plus, and it beat out the competition two years running in *PC Magazine*.

That's because TaxAct offers easy navigational tools and plenty of help, such as "TaxTutor Guidance." It's also the least expensive of all the major online tax programs. And, for a few extra bucks, you can make unlimited calls to a tax expert during filing season. It almost makes doing your taxes a sweet deal. Almost.

Pay with your phone, leave the cash at home

Your smartphone is now a wallet. With mobile apps like PayPal, Android Pay, and Apple Pay, paying for stuff has never been easier.

- Paypal stores your bank and credit card information right in the app. Simply swipe your phone when making a purchase, or select PayPal at the register and pay by logging in to your PayPal account. You can even check your balance or send money to a friend in need right from your phone. PayPal protects your account information, and payments go through only after you enter either a secret PIN or your username and password.
- Android Pay and Apple Pay work the same way but on Android and Apple devices, respectively. Unlike PayPal, Android and Apple Pay do not store your bank or credit card information on your phone. What's more, you can tell Apple Pay to require Touch ID, or fingerprint verification, before completing a purchase.

Not all stores accept PayPal, and not all phones have a Near Field Communication (NFC) chip, which is needed for Android and Apple Pay. Ask your mobile carrier if your phone does.

Hassle-free way to send and receive cash fast

Does someone owe you money? Now it can be put in your bank account almost instantly, without anything changing hands! The app Square Cash could make writing or cashing checks obsolete. It makes sending and receiving money quick and easy — even if that friend, relative, or charity doesn't have a Square Cash account.

You can send and receive payments by email or by downloading the app from Google Play or the iTunes App Store. You don't need a Square Cash account, but you do need a debit card (not a credit card) backed by Mastercard or Visa and tied to a bank account. That's what allows Square Cash to transfer money from one person's account to the other's. Follow these steps to send funds via email.

1. Open your email program and start a new email.
2. Type your recipient's email address in the *To* field. In the *CC* (carbon copy) field, type "cash@square.com." This sends a copy of the email to Square Cash, telling it that you want to transfer money. Type the amount of money in the *Subject* line, such as $10, and send the email.
3. If you don't have a Square Cash account, you will get an email reply from Square Cash asking for your debit card number. Type it in and click *Send*.
4. Your recipient will also get an email from Square Cash asking them to fill out a form with their debit card number. Once you both give your card numbers to Square Cash, the money will be transferred within two days.

Sending money using the app is even simpler. Download it, and the app will help you create a Square Cash account the first time you open it. Follow the app's instructions for sending and receiving money. Square Cash is free for personal use, and the money gets deposited directly into your bank account.

This service is only as safe as your email is secure, though. Be sure to lock your email account behind a strong password, and check your bank account regularly for unusual activity. For an added layer of protection, go ahead and create a Square Cash account. You can set up your account so that Square Cash asks for the three-digit security code on the back of your debit card before it transfers any money out of your bank account — a code that hackers aren't likely to have.

Simplify splitting your dinner tab with friends

Splitting a dinner tab has never been easier. The Tab app will calculate the totals for everyone at the table and include the tax and tip! Simply snap a picture of the receipt and let each person tap their items. Tab will even divvy up the cost of shared food. Search your phone's app store for "tab" and look for *Tab — The simple bill splitter* or *Tab — The simple way to split a bill.*

Spend less, save more with this savvy app

Stop spending more money than you need while shopping. Take ShopSavvy with you. This handy app delivers instant price comparisons while you shop. Retailers hate it, but you will love the money you save.

Download the app on your smartphone or tablet. Open it up, scan an item's barcode with your device's camera and — shazam! — ShopSavvy shows you current prices at both online and brick-and-mortar stores.

But wait, there's more. Check the ShopSavvy home screen for featured sales online and in local stores. Want to stay in your neighborhood? No problem. Just tap the *Nearby Sales* button at the top. You'll only see sales in your area. No more sorting through Sunday's sales fliers.

Make grocery shopping cheaper and easier

Next time you head to the supermarket, take GroceryiQ with you. This helpful little app will put together your grocery list, organize it by aisle to save you time in the store, and match items with related coupons.

You can add groceries to your list by typing in each item along with the brand and even the flavor. Typing is so last year, though. Tap the microphone icon in the upper right corner and dictate your items without lifting a finger. GroceryiQ will show you a list of matching foods, along with pictures. Tap the one you want to add it to your list.

Do you still have the now-empty package? Scan its barcode with your phone's camera to add it to your list. As you build your list, GroceryiQ will find related coupons that you can print and take to the store. Or you can browse the app's gallery of more than 200 coupons. Grocery shopping will never be the same again.

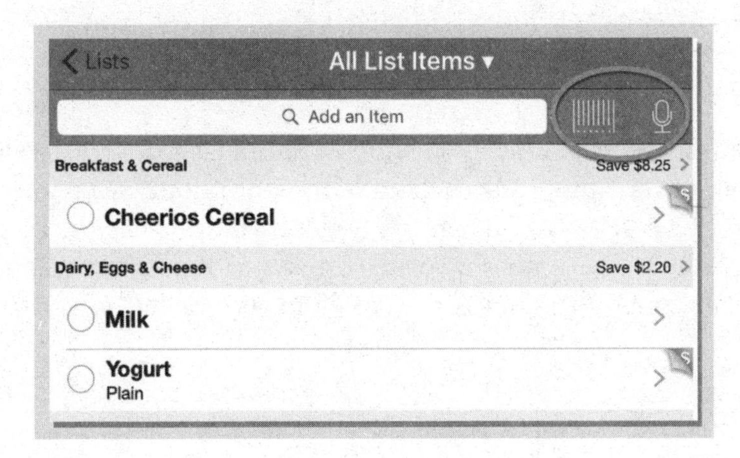

Check this site before you buy a used car

Before you buy that used car, be sure you don't get stuck with a recall issue. Here's the easy way to do it. Go to *Safercar.gov* and click the *Search for Recalls by VIN* button. Then enter the car's vehicle identification number (VIN). You'll find out if the vehicle has any outstanding safety recalls from the past 15 years that haven't been repaired. No more wondering if you're buying a lemon.

The secret to saving money on just about anything

Don't you just love paying full price for everything? Nope, didn't think so. That's why you need the app Shopular. Pick your favorite stores from a list of hundreds, including:

- big box stores like Target and WalMart.
- supermarkets like Albertsons and Kroger.
- home improvement shops like Ace Hardware, Lowe's, and Home Depot.
- clothing stores like Gap and Old Navy.
- department stores like Kohl's and Macy's.
- sporting goods dealers like Dick's and Sports Authority.
- and even eateries like Subway and Starbucks.

You get the picture. Once you choose, Shopular loads local sales and coupons for these stores directly to your phone or tablet. Thanks to its geolocation feature, it alerts you to deals the moment you walk into one of your favorite stores. Just hand your phone with your coupon to the cashier. You'll never pay full price again!

Save and send your favorite deals

Tap the heart next to a deal or product you like in Shopular, and the app will save it for you to view later. To share a deal with a friend, tap the paper airplane next to the heart. This attaches the Shopular deal to a text message you can send. (See example on next page.)

Sell stuff safely on Craigslist

Want to sell your stuff without the hassle of a yard sale? Give *Craigslist.com* a try. You can advertise items for sale online without paying a fee. Plus, Craigslist reaches more buyers than your local paper's classifieds. But use it with caution. Here are three things you should never share with people online. Use these tips to protect yourself.

- Your email address. Keep your email anonymous by selecting *CL mail relay (recommended)* under *Contact Info* when you create your ad, and only list your first name.

- Highly personal information. Keep your identity safe on Craigslist. Don't share your bank account number, birth date, or social security number. Otherwise, you could easily become a victim of identity theft.
- Your home address. Always meet buyers in a public place, take a friend, and have your cellphone handy in case something goes awry.

No-stress way to remember your passwords

Need help managing all your passwords? A password management program is the way to go.

The beauty of a password manager is its simplicity. You just have to remember one password. Ever. That password unlocks your chosen password management program, and it does the rest. The program will generate a different, uncrackable password for everything you do on your computer. It's a great way to keep your passwords secure from snoopers and scammers.

You can choose from literally dozens of password manager applications. One popular program is called Sticky Password. Go to *stickypassword.com*, and install the app on your computer, tablet, or phone. It works on Android and Apple devices as well as personal computers.

The first time you open it, Sticky will ask you to create a new account. Enter your email address, then create a strong Master Password, one that you can remember but that would be hard for a thief to guess. Sticky suggests making it at least eight characters long with a mix of numbers, punctuation marks, and upper- and lowercase letters.

- The first time you log in to your email, bank, or other online account, Sticky Password will offer to remember your login information. Say yes, and from then on the app will automatically fill in your username and password every time you visit that website.
- When you create new accounts online, use Sticky to generate a super strong password and remember it for you.
- This app can also store your credit card information, address, and other personal details and use them to fill in online forms for you. No more tediously typing in your mailing address or credit card number when you place an order!

You can access your Sticky Passwords from all of your devices. Plus, Sticky's encryption feature makes it super secure, so you never get hacked. To get the free version, go to *StickyPassword.com*. Upgrade to the paid, premium version, and part of the proceeds go toward saving manatees around the world. You'll get password protection and a warm, fuzzy feeling for doing something good.

Find words fast with this Dictionary app

Stop looking up words in a big dictionary. Now you can do it anytime — on your phone, tablet, or computer — with *Dictionary.com*.

This website and its related app will define words for you, use them in sample sentences, show you synonyms and antonyms, and give you the brief history of a word. You can even look up acronyms like ROFL (rolling on the floor laughing). No more wracking your brain for something to say.

5 top sites for finding your family roots

There's only one way to find out if your great-great-grandfather was a hero in the Spanish-American War — dig into your family roots. It's easy when you let the power of the Internet do the work. Search for genealogical websites, though, and you'll find hundreds. So what's the best online genealogy site for you? Check this list to find out.

Website	Cost	Special features
Ancestry.com	Free 14-day trial, then monthly or 6-month membership fee	Helps you discover your immigrant ancestors and explore extensive marriage, birth, and death records
MyHeritage.com	Free for a basic plan; monthly fee for enhanced plans	Helps you build your family tree and search international records
FamilySearch.org	Free	Lets you upload photos, documents, and audio records to create a family album
FindAGrave.com	Free	Shows the final resting place of more than 138 million people
WikiTree.com	Free	Lets you contribute to one giant family tree that connects you with the rest of the world

Picking paint colors just got simple

Don't stress out about choosing the right paint color for your home project. Let the Behr ColorSmart app help you. This smartphone app works like an interior designer, helping you with everything from color coordinating your rooms to calculating how much paint you will need.

With the Color Match tool, you can even snap a photo of an object — say, a bright yellow rose or a navy blue blanket — and the app will not only find a matching paint color, it will also show you several complimentary colors.

What's more, the app comes with a preloaded gallery of room photos, such as a contemporary kitchen, a country-casual bedroom, and a traditional living room. The coolest part of this feature? Tap on the room that most resembles the one you want to paint in your own house. Several swatches will appear with four coordinating colors plus the photo of the room. Tap a swatch, tap one of the colors, and then tap a wall in the room. The app will "paint" the wall for you so you can see how it will look in your own home.

One app is all you need to organize your life

Stop what you're doing right now and download Evernote onto your phone, tablet, or computer. Then get ready to organize everything in your life, from making a shopping list to setting weekly health and fitness goals.

You can create plain old lists, but where's the fun in that? You want the feeling of accomplishment that comes with checking off each item. So turn that ordinary Evernote list into an esteem-boosting list of checked-off achievements. Just open the app and follow these instructions.

- On your computer, click + New Note to create a new list. Type a title and press return. Look at the toolbar at the top of your note and click the checkbox. Now every time you make a hard Return, a box will appear.

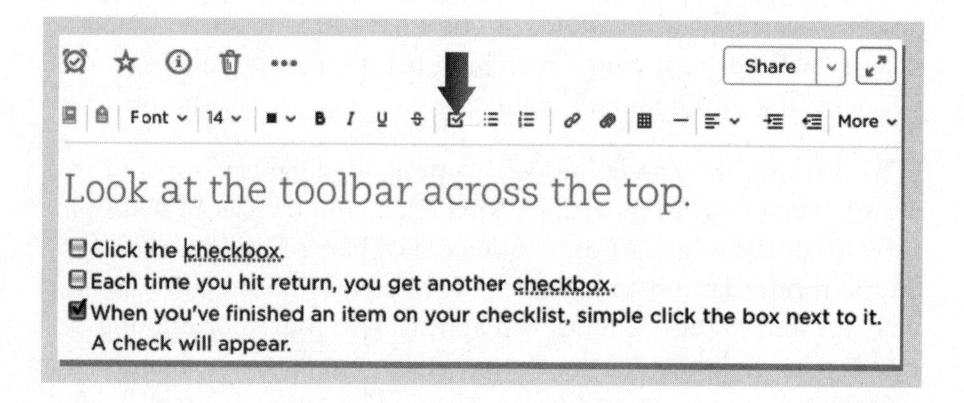

- On your phone and tablet, tap the List circle to open a new check list, and start typing. Each new line will begin with a checkbox.

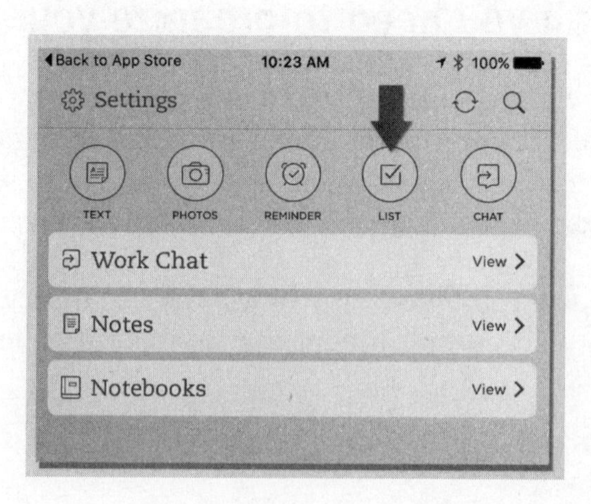

Simple yet versatile note taking

You've heard the phrase "there's more than one way to skin a cat." With Evernote, there's more than one way to create a new note. Open the Evernote app on your smartphone or tablet, then tap the tiny elephant in the upper left corner to open your settings and toolbar.

- Tap *Text*, and a new, blank note appears. Simply start typing or choose one of the options on the note's toolbar.
- Tap *Photos* to snap a picture of a receipt or business card, for instance, or to select a picture from your device's photo gallery.
- Tap *Reminder* to open a small text box. Think Post-it note. Scribble your reminder and set it for a specific date and time.
- Tap *List*, and your new note automatically becomes a checklist.
- Tap *Chat* to send an instant message to someone in your Evernote contact list.

Flip for a magazine that's made just for you

Imagine flipping through a magazine in which every single article interests you. Sound impossible? It's not with Flipboard, a brilliantly designed digital journal available online or as an app for your phone or tablet.

Flipboard has hundreds of topics for you to follow like investing, book reviews, home decorating, technology, adventure travel, and Asian recipes, just to name a few. Pick the topics that interest you, and Flipboard pulls together top stories on your favorite subjects from all over the world.

You get gorgeous color photos or videos with each article, too. It's perfect for filling the lulls in your life, whether you're flying coast to coast or, worse, waiting at your doctor's office.

Send, receive, and print faxes for free — without a fax machine

You don't need a fax machine anymore to send or receive a fax. Sign up for an online fax service such as *FaxZero.com*. It lets you send and receive five free faxes a day, up to three pages each, from your home computer. All you need is an email address and an Internet connection. That's it. For a small fee, you can send longer faxes, up to 25 pages, directly from your computer.

If someone needs to send you a fax, turn to *MyFax.com*. You and the sender must register your email addresses with this service. When someone sends you a fax, it will appear as email. The first 30 days of *MyFax.com* are free. After that, you'll pay a small monthly fee to keep the service.

Enjoy Web articles and videos without an Internet connection

Tuck articles and videos in your Pocket and pull them out later when you have time to enjoy them. No, not that place in your jeans where you stash your keys and spare change. Pocket is an app that lets you store articles and videos from online directly to your phone, tablet, or computer. So if you don't have time to read that long news article now or to watch your son's latest YouTube video, just put it in your Pocket and save it for later.

Install the app on your device or look for the Pocket button on your Web browser's toolbar. Tap the button to save that item.

Pocket syncs across all your devices, and you can open your saved items at any time, even without an Internet connection. It's sure easier than digging for loose change.

Learn almost anything from the privacy of your home

Imagine taking a class in your pajamas with your first cup of coffee at your side. Learn how to use new programs from the comfort of your own home with these two options.

- Teach yourself to use the computer with *YouTube.com*, where you can search for anything from how to take a screenshot to troubleshooting an error message. Most YouTube videos offer step-by-step instructions that anyone can follow.
- Take professional courses by signing up for a membership at *Lynda.com*. This site offers more than 4,000 online courses, including lots of computer classes, for a monthly or yearly fee. That's not all, though. You can also take classes in fields like business, marketing, education, technology, and languages, all taught by industry experts.

Find a parking spot without going crazy

Stop driving in circles looking for a place to park. SpotHero will find you a spot in a parking garage in more than a dozen big cities nationwide, at a discount.

Search ahead of time from your computer at *SpotHero.com*, or download the app onto your phone and find parking on the go. Type in a landmark or address, and the app will display a map with multiple garages, along with prices for each one. Tap a garage, and you'll see its hours of operation, restrictions and amenities, and a button to reserve and pay for your spot.

Keep track of any flight's arrival

No more circling the airport pickup line! FlightAware lets you track a flight's whole trip and check on its safe arrival before you pick up your loved one. This free app shows you loads of information, from the speed, altitude, and duration of a flight to the plane's location on a map in real time. Download the app to your device, or visit *FlightAware.com* to see the flight information on your computer.

Save 70 percent on hotels with a few simple swipes

It's easy to save money on hotels when you have the right apps on your phone. Start with HotelTonight, an app that lets you book a room with three taps and a swipe. It can save you up to 70 percent off rooms at hotels that range from basic to luxe. Why? Because HotelTonight helps fill up rooms that would otherwise go unsold.

Roomlia, another app, works in much the same way. A few swipes on your smartphone, and you've booked yourself a great hotel deal. Both apps specialize in last-minute deals, so you can't book more than seven days in advance, but you can book as many as five nights in a row.

Outsmart traffic with Waze

There's an easy way to find the best route and beat the traffic to your destination. It's an app called Waze, and it combines social media with traffic navigation. Other drivers post traffic alerts on Waze in real time, so you learn about accidents up ahead or road work on the way immediately. Waze will even show you alternate routes. Plus, that network of drivers warns you about nearby speed traps and tells you where to find the cheapest gas. The map even shows cartoon-like icons of police officers and where they're parked.

The biggest problem with using Waze is getting distracted while driving. Pull over if you need to study a map carefully or update traffic conditions on your route. Waze promotes itself as a community of drivers working together to outsmart traffic. It's true, so the next time you're on the road, let Waze lead the way.

Get a ride to almost anywhere with one tap

Need a ride somewhere? Try Uber, a cheap and convenient alternative to taxis. When you open the Uber app on your phone, it will locate you, show you drivers nearby, and tell you how long before a driver can pick you up. You can also see the driver's photo, phone number, and rating. One tap, and your driver will be on his way to get you. No need to exchange any cash. Link a credit card to your Uber account, and Uber will charge the fare to your card. Uber easy.

The secret to getting your flight for a cheaper price

It's the app with the funny name — Yapta — and it's nothing short of "your amazing personal travel assistant." Here's how it works.

Yapta tracks airfares and sends you an alert when a price drops. Then you decide if you want to purchase the fare or wait to see if it drops even more.

So what happens if you've already bought your ticket? Yapta will send you an alert letting you know if and when you can get a refund check or credit for the difference from the airline. Check *Yapta.com* to see which airlines offer refunds.

Get free Wi-Fi wherever you go

You're going out of town, but that doesn't have to mean you're disconnected. You can still stay on top of news, check email, and share photos from your trip. Lots of businesses provide free Wi-Fi access to keep their customers happy. And with WiFi Map, it's a cinch to find free Wi-Fi access wherever you are.

This app shows you a list of places with free W-Fi access, along with a corresponding map. Simply type in your location or allow the app to check your device's location.

WiFi Map also gives you the passwords for public Internet hotspots, so you can log in to the Internet from anywhere in the world. The basic version is free. For a few dollars, the Pro version gives you access to a map when you are offline that directs you to free Wi-Fi.

Facebook, Twitter & Pinterest

Fine-tune your sharing

8 stupid things smart people do online — and how to avoid them

Social media keeps you on your toes. If you're not posting pictures of your latest family reunion on Facebook, you're showing off your new grandchild on Instagram. But all this sharing comes with risks. Want to be safe? Find out what not to do on Facebook, Twitter, LinkedIn, Google Plus+, Instagram, and other social media sites. Here's what you need to know to be a savvy user.

Postpone posting vacation photos. If there's one thing you should never, ever do on Facebook, it's announce that you are on vacation while you are still on vacation. Don't post status updates from the airport or share photos of your trip until you return — unless, of course, you want criminals to break into your house while you're gone.

Stop checking in while you're out. Telling the world you're at a concert, at the movies, or out shopping is like telling thieves and stalkers, "Hey, I'm not home. Help yourselves!" So stop "checking in." Keep your whereabouts to yourself as much as possible.

Keep your cool. Everyone needs to vent, but social media is not the best outlet. Angry or critical posts tend to bite you in the behind later on. Plus, friends may change their opinion of you if you go off on frequent rants.

Don't blast your boss or your coworkers. Even if your colleagues are not on your list of Facebook Friends or Twitter Followers, negative comments about your job could come back to haunt you. Your current supervisor may hear about your rants through the grapevine. Potential employers may get turned off by your tirades over your current work situation. Play it safe and separate your social-media life from your professional one.

Curb your selfie habit. Think twice before posting a photo of yourself on Facebook. Hackers can use facial-recognition software to gain your personal information, maybe even your Social Security number, just from pictures of you on social media. That's what a study out of Carnegie Mellon University found. Scary. Minimize the number of photos that you post of yourself and are tagged in.

Keep this one thing a secret. Ever use your birth date as a password? You're not alone, and hackers know it. Keep your birth date to yourself. And if you absolutely must share it, only reveal the month and day, not the year.

Beware how you watch videos. There's a sneaky way that watching a "harmless" video on Facebook or Twitter could lead to a malware infection on your computer. It happens when you click on the video and get a message telling you that you need to download a special media player to watch it. Be smart and avoid that trap. Search for the same video on YouTube and watch it there, instead.

Be picky about your Friends. Check your list of Friends and people you follow from time to time. Drop any you don't know well, and keep the ones you trust. On Facebook, don't confirm Friend requests from people you don't recognize. They could be thieves or scammers targeting you.

Restrict who sees your Facebook posts

You have complete control over who sees your Facebook posts. Want to share something with family only? No problem. Just with close friends? Sure thing. You can even share your post with everyone on Facebook. The *Who should see this?* button makes it all possible.

Look for a button labeled *Friends* or *Public* at the bottom of your status update. Click this, and Facebook will let you pick who can see this post. You can share it with all of your Facebook friends by choosing *Friends*; with everyone on Facebook by clicking *Public*; or with select people by clicking More Options > Custom, and typing their names into the *Share with* box.

All of these options come with one warning — Facebook remembers which group you choose, and it will only share your next post with the same group. So remember to double-check who is set to see your post the next time you share something.

Control what you see in your Facebook News Feed

Tired of scrolling through a lot of nonsense on your Facebook feed? You're not alone. Here's how to gain more control and become the boss of your own account.

- Click the drop-down arrow beside *News Feed* on the left side of your Home page. Then choose *Most Recent.* This tells Facebook you want to see your friends' new posts at the top of the list, rather than the posts that Facebook wants you to see.

- If you no longer see a friend's posts in your News Feed, visit his page and click *Like* on a few of his posts, then leave a comment or two. This should trigger his posts to start showing up again in your feed.

- What can you do about that annoying, over-sharing friend or family member on Facebook? Don't unfriend her. Just click the arrow in the upper-right corner of the person's post. Choose *Unfollow*, and you'll stop seeing her posts.

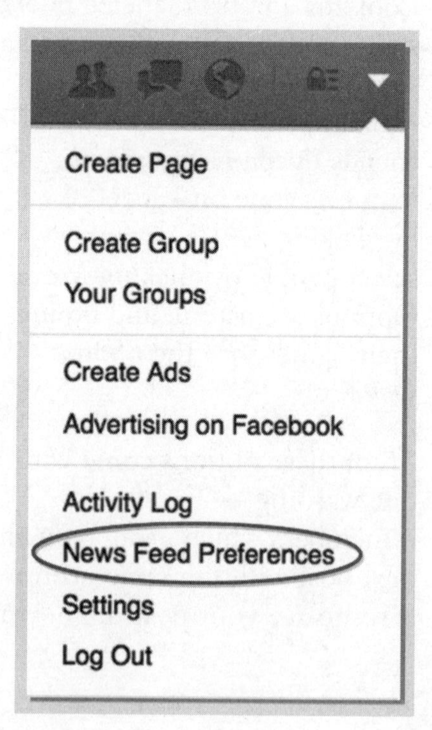

You can also click the arrow in the upper-right corner of your Home page, next to the padlock-shaped icon. Click *News Feed Preferences* in the menu that opens to start customizing what you see in your News Feed.

Stop thieves from hijacking your Facebook account

There are lots of sneaky people in the world, and some of them want to log in to your Facebook account without you knowing it. Here's the clever way to keep someone from doing that.

1. Click the arrow in the top-right corner of your page, beside the padlock icon.
2. In the menu that appears, click *Settings*, then click *Security* on the left side of the window.

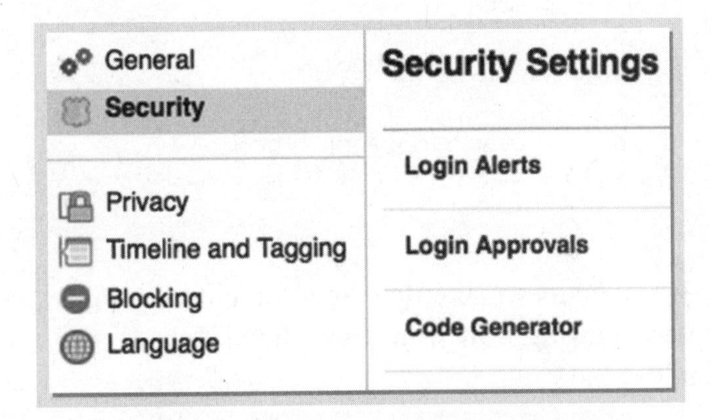

3. Click *Login Approvals* in the center, then click to place a check mark in the box beside *Require a security code to access my account from unknown browsers.*

Facebook will walk you through the steps to create a security code, which it will send as a text message to your cellphone. No one else will be able to log in to your account from a strange computer unless they have that code.

Surf Facebook faster with these shortcuts

Forget about clicking! The best way to get around Facebook is hidden in plain sight on your keyboard. No need to use Shift, Ctrl, or any other fancy keys. Simply press one of these ordinary letters.

Press this	To do this
j	Scroll down through your News Feed
k	Scroll up through your News Feed
p	Update your status
/	Search Facebook
q	Search for people on Chat

These next shortcuts work only when you click on a post, not when you are viewing it in your News Feed.

Press this	To do this
l	Like or unlike the post
c	Comment on the post
s	Share the post

Get where you need to go without using your mouse

Some Facebook shortcuts are a little more complex — you have to press three keys on your keyboard instead of only one. Once you get comfy with the single-key shortcuts, take things up a notch. Learn how to bebop around Facebook like an old pro with these key combinations. They work in most, but not all, Web browsers. Just follow the commands for your type of computer.

To go here	Press this	
	On a Mac	On a PC
Facebook's Help Center	Ctrl+Option+0	Shift+Alt+0
Your Home page	Ctrl+Option+1	Shift+Alt+1
Your Profile page	Ctrl+Option+2	Shift+Alt+2
Friend Requests and people you may know	Ctrl+Option+3	Shift+Alt+3
Your Messages Inbox	Ctrl+Option+4	Shift+Alt+4
Your Notifications	Ctrl+Option+5	Shift+Alt+5
Account Settings	Ctrl+Option+6	Shift+Alt+6
Your Activity Log	Ctrl+Option+7	Shift+Alt+7
About Facebook	Ctrl+Option+8	Shift+Alt+8
Facebook's Terms and Policies	Ctrl+Option+9	Shift+Alt+9
Start a new Message	Ctrl+Option+m	Shift+Alt+m

Save that fascinating Facebook post for later

You really want to watch the cute puppy video your daughter shared on Facebook, but you don't have time at the moment. Fortunately, you can save it for later with one of Facebook's handy features.

Click the little arrow in the top-right corner of any post you want to save. From the drop-down menu, click *Save post, Save link*, or *Save video*. This will store the post in your Saved folder.

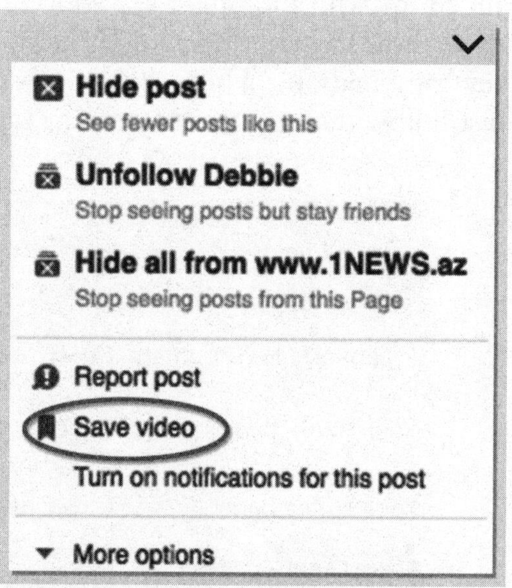

When you're ready to read that article or watch that video, go to your Home page and click *Saved* on the left side, under *Favorites*. Then click one of your saved posts to view it.

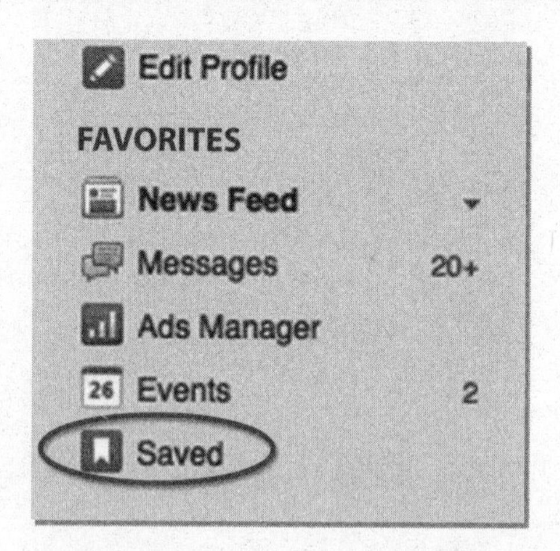

You can save them for as long as you like and delete them when you're ready. Just click the little arrow in the upper-right corner of the post again and choose *Unsave*.

Take control of the ads in your News Feed

Facebook is free, but in return you have to put up with lots of annoying ads in your News Feed. Or do you? Here's a simple way to cut the clutter in your Facebook feed.

1. Click the little arrow in the top-right corner of the ad. Click *Hide ad*, and you'll see fewer ads like it in the future.
2. Better yet, click *Why am I seeing this?* You'll get a basic answer and gain access to another useful option called *Manage Your Ad Preferences.*

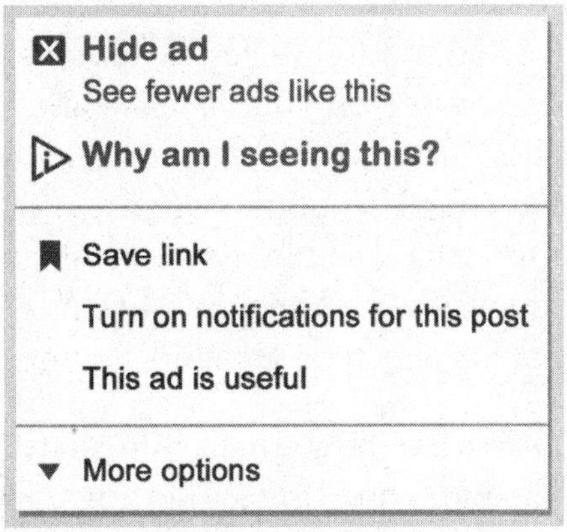

Click this, and you'll see a long list of categories like *Family and Relationships, Lifestyle and Culture*, and *Shopping and Fashion.* Clicking one of these topics opens a list of subtopics.

For instance, under *Family and Relationships* you may find *Childbirth, Home*, and *Motherhood.* Hover your mouse pointer over any subtopic and click the *X* on the right to get rid of it. You will no longer see ads pertaining to that topic.

These steps won't stamp out Facebook ads completely, but they will give you more control over what you see.

Learn what strangers can see about you on Facebook

You know what your Facebook looks like when you log in to your account, but what do friends and strangers see when they view your page? It's easy to find out.

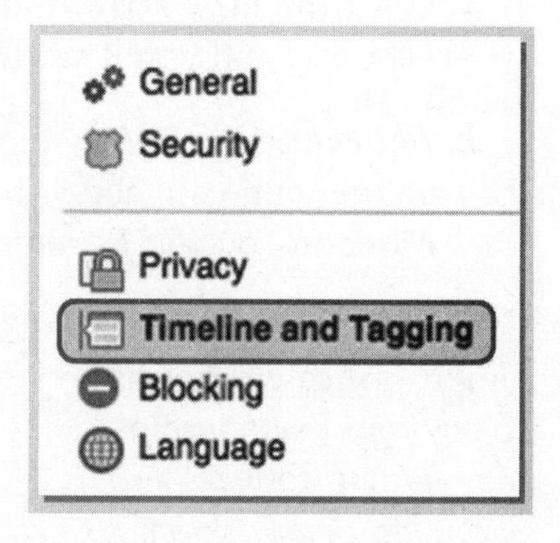

On your computer, click the little arrow in the top-right corner of your Facebook page, next to the padlock icon. Click *Settings* then *Timeline and Tagging.*

In the center of the window, look for the phrase *Review what other people see on your timeline* and click *View As.* You'll immediately see your Facebook page the way the general public does. To see it the way your Facebook Friends see it, click *View as a specific person* and type a friend's name into the Search box.

Don't like how much information people can see about you? Change your Privacy settings. Exit the "View As" feature by clicking the *X* in the upper-left corner of the window. From your Home page, click the padlock icon in the upper-right corner and choose *Who can see my stuff?*

Forgot to log out? Take action fast

D'oh! You were at the library "just checking Facebook for a minute," and you forgot to log out of your account. Well, you're in luck. You can log out of Facebook from any device after the fact, and keep potential identity thieves from taking advantage of your mistake.

Click the arrow in the upper-right corner of your Facebook page, then click *Settings*. Next, click *Security* on the left side of your window.

Look for the category *Where you're logged in* in the center of the window, near the bottom, and click on it. This opens a list of all the computers, mobile phones, tablets, and other devices that you have used to open Facebook recently. Scroll through the list and, if anything looks suspicious, click *End Activity* to log out on that device.

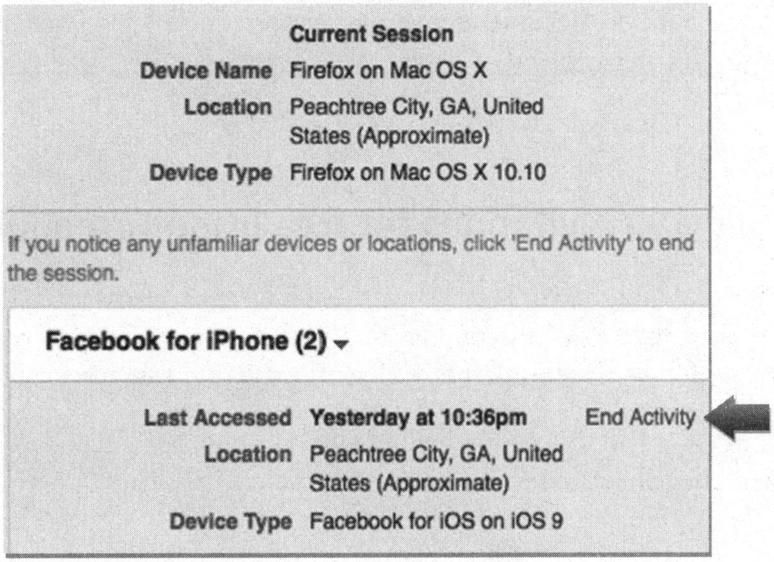

Current Session
Device Name Firefox on Mac OS X
Location Peachtree City, GA, United States (Approximate)
Device Type Firefox on Mac OS X 10.10

If you notice any unfamiliar devices or locations, click 'End Activity' to end the session.

Facebook for iPhone (2)

Last Accessed Yesterday at 10:36pm End Activity
Location Peachtree City, GA, United States (Approximate)
Device Type Facebook for iOS on iOS 9

Limit who can see old, embarrassing posts

Sharing an embarrassing photo on Facebook may seem like a good idea at the time. Leaving it on your Timeline for the world to see forever and ever? Not so much. Thank goodness you can hide old posts.

To hide a lot of old posts at once, click the padlock in the top-right corner of your Facebook page. Click *See More Settings* at the bottom of the menu to open the *Privacy Settings and Tools* page. Under *Who can see my stuff?*, click *Limit Past Posts* on the right. Then click the buttons *Limit Old Posts* and *Confirm*. Now, the only your Facebook friends will be able to see these posts — not strangers, not your boss, and not the general public.

You can also hide individual posts. Just click the arrow in the upper-right corner of the post and choose *Hide from Timeline*. Poof! Problem solved.

Send 150 invitations for free using Facebook

Planning a big party is fun, until you have to address and stamp 150 invitations. Well, you can finally put away your stamps and envelopes — Facebook lets you send private messages for free.

Click on a friend in your Friends list to open her Facebook page, then click the *Message* button on her cover photo.

When the message window pops open, click the people icon to add more friends to the chat. You can message up to 150 people at once.

What if you're on the receiving end of a group message that you don't want to be part of? Just drop out of the chat. From your phone or tablet, open the message and tap the info icon in the upper-right corner.

- Then, on iPhones and iPads, scroll to the bottom of the screen and tap *Leave Group*.
- On Android phones and tablets, tap the three dots in the top-right corner of the screen, then tap *Leave Group*.

Or, from your computer, open the message, click the gear-shaped icon to the right of the group's name, and then click *Leave Conversation*.

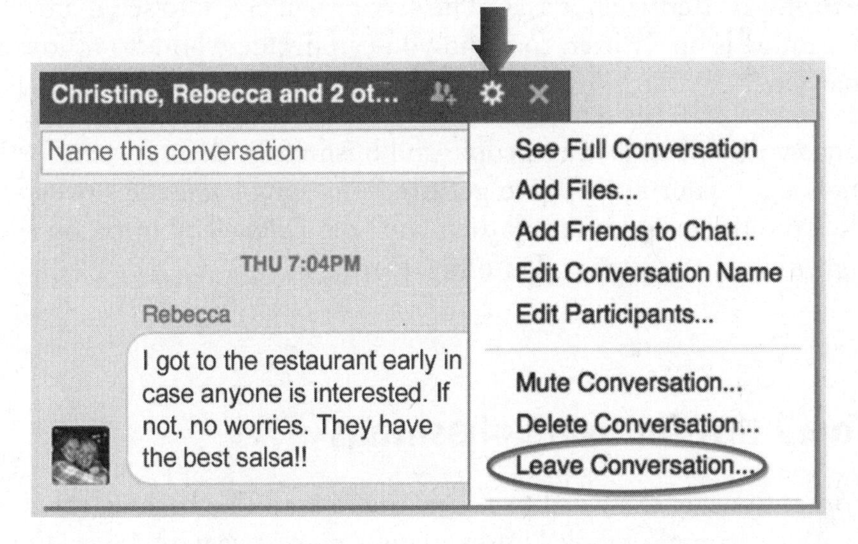

Figure out who to follow on Twitter

You've heard people talking about Twitter, but figuring out how to use it and who to follow is just too much trouble, right? Not if you let Twitter help. It suggests all sorts of people to follow when you first sign up.

After creating your username, Twitter will ask about your interests to get a feel for what you'd like to see. You can pick from News, Music, Sports, and Fashion, to name a few. Simply click your favorite categories. Based on your picks, Twitter will suggest specific people and organizations you might enjoy following, such as Taylor Swift, the Atlanta Braves, or *Vanity Fair* magazine. Put a check mark next to the ones you want to follow, and click *Continue*.

You can also follow friends and relatives who have Twitter accounts. While setting up your account, simply share your email Contact list with Twitter when it asks. The service will check to see if anyone you know is on Twitter, too, and you can decide whom to follow at that time.

Once you start following people and businesses, their "tweets" will show up in your feed. If you get tired of seeing someone's tweets, click the person's name and then click the *Following* button on the right side of the page to Unfollow them.

Top 3 tips for using #hashtags

Forget the egg, the word tweet, or the @ key. The hashtag (#) is Twitter's iconic symbol. Unfortunately, many tweeters don't know what to do with it. Here's a quick, crash course.

- Use a hashtag to find topics that interest you on Twitter. Type a name or phrase in the *Search Twitter* box with # in front of it, like #carrieunderwood or #fantasyfootball. You'll see a number of feeds you can follow. Just click to choose one you like.
- If you want to follow an event, type # and the event's name in the *Search Twitter* box — for instance, #presidentialdebates or #academyawards. You'll get the latest news on the event, plus see real-time updates from people who are there.

- Create buzz around something you want to promote by using #. Say you want to spread the word about your store's grand opening or a free donut giveaway. Send a tweet from your account with a phrase such as #grandopening or #free-donuts, along with a brief description. Your followers will thank you.

How your #Hashtags can lead to #BigDiscounts

Cashing in on Facebook, Twitter, and Instagram is easy when you use hashtags. Type a hashtag in front of key discount words, like #coupon, #BOGO, and #giveaway, into the Search box on your favorite social media site. Add a brand name, such as "Clinique #giveaway," if you're looking for something specific. You'll get a long list of money-saving deals. #Enjoyyoursavings!

Hit mute to stop obnoxious tweets

Tired of someone's constant tweets? Hit the mute button — literally. This feature lets you stop their tweets from cluttering up your Twitter feed without unfollowing them. Plus, if you follow someone who retweets posts you don't like, you can mute the annoying retweets but still see original tweets from the person you follow.

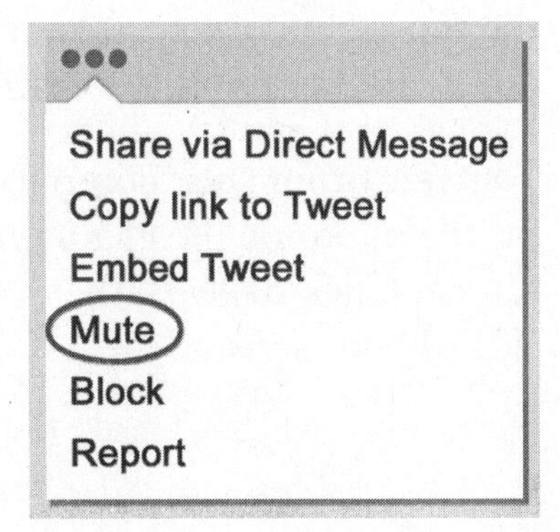

- From your computer, click the ellipsis dots (...) at the bottom of a tweet, then click *Mute* in the menu that appears.

- On a smartphone or tablet, press and hold your finger on the tweet for a moment, then lift your finger. Tap *Mute* in the menu that appears.

Protect your privacy on social media

Most people tailor their conversation to present company, right? You can do the same thing with social media posts — especially if you don't want strangers to scan your posts and steal private info. After all, you don't want thieves to see your post to friends about an upcoming vacation and rob your house while you're gone. Here's how to limit who sees what on Facebook and Twitter.

Facebook. On your computer, open Facebook and click the padlock icon in the upper-right corner of the window. Then click *Who can see my stuff?* From here, you can change who sees your posts.

On a phone or tablet, open the Facebook app and tap the ≡ icon. It's in the bottom-right corner of iPhone and iPad screens, and in the top-right corner of Android phones and tablets. Then scroll down and tap *Privacy Shortcuts* to tweak your settings.

Twitter. From your computer, log in to Twitter and click your profile photo or the egg icon in the upper-right corner of your screen. Click *Settings*, then click *Security and privacy* on the left side of your screen.

On an Apple device, tap the *Me* icon at the bottom. Then tap the gear-shaped icon, go to Settings, and tap your account name.

On an Android device, tap the three dots in the upper-right corner of your screen. Then tap Settings > Privacy and content.

At the very least, consider putting a check mark in the box beside *Protect my Tweets*, and tap to double-check your settings under *Who can tag me in photos.*

Use Lists to keep your Twitter feed tidy

Scrolling through your Twitter feed can be frustrating, especially if you follow people who tweet constantly. Thankfully, you can organize all the people and things you follow into custom lists. Maybe you want to see all the political tweets in one place and all the cooking tweets in another. Just group them into separate lists.

From your computer's Web browser, log in to Twitter, click your profile picture in the top-right corner of your page, then click *Lists*. In the next window, click the *Create new list* button. Give it a name and brief description, decide if you want it to be *Public* or *Private*, then click *Save list.*

Your list starts out empty, even if you currently follow 50 people on Twitter. You'll have to add them individually to any list you create, which you can do in two ways.

- Click one of your lists on the left and search for people or organizations in the Search box under *Find people to add to your list.*

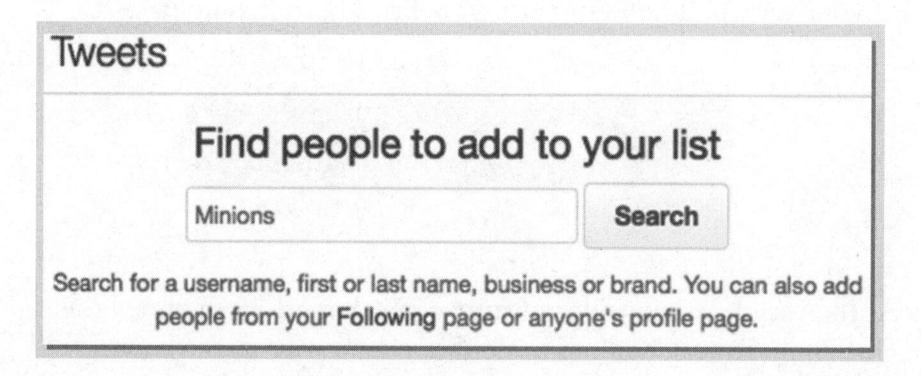

- Search for people from your Twitter Home page. Type a name into the *Search Twitter* box at the top and click a result to open that page. Then click the gear-shaped icon and choose *Add or remove from lists.*

From your Android device, open Twitter and tap the three dots in the top-right corner of your screen. Tap *View Lists* then the plus sign (+) to create a new list. Name it, describe it, make it private if you want, and tap *SAVE.*

Get back to your home page by tapping the bird icon along the top. Once there, tap the magnifying glass to search for people to add to your list. Type a name into the *Search Twitter* field and tap one of the results to open that user's profile page. Finally, tap the three dots in the top-right corner and choose *Add to List.*

From your iPhone or iPad, open Twitter and tap *Me* at the bottom of your screen. Tap the gear-shaped icon, then *View Lists.* Tap the plus sign (+) to create a new list, then name it and tap *Save.* You can search for people to add to your list right away on the next *Members* screen.

Or do it later — just tap the magnifying glass in the top-right corner of your Home page and start typing. Tap one of the results to open that profile page, tap the gear icon, then tap *Add/remove from lists.*

Share great tweets with just 2 clicks

Read a tweet you just love and want to share it with your legion of Twitter followers? Nothing simpler. Sharing a tweet is called retweeting (RT), and it's a great way to befriend the people you follow on Twitter. Simply click or tap the RT symbol on your favorite tweet.

Next, you have a chance to add your own witty comment to the original tweet. On a computer, type your comment and click *Tweet.* On a phone or tablet, you'll see two options after tapping the RT symbol — *Retweet* and *Quote Tweet.* If you don't want to add a comment, just tap *Retweet.* But if you want to add a short statement, tap *Quote Tweet.*

Secrets to searching for specific tweets

Twitter's Advanced Search tool makes it easy to track a breaking-news story as it unfolds, or to find tweets about a specific topic without scrolling through thousands of tweets.

Let's say a python escaped from the nearby zoo, and you want to know if any Twitter users have spotted it — you know, in case you need to be extra careful on your way to the mailbox.

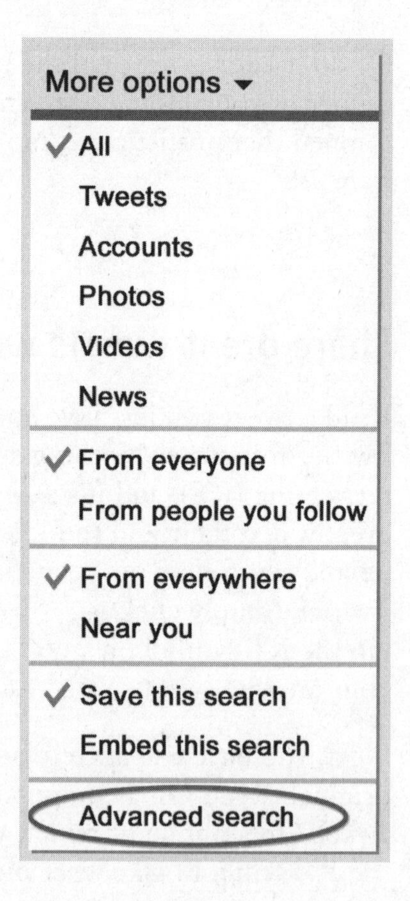

1. Sign in to Twitter from your computer and type some key words into the *Search Twitter* box, such as "escape snake zoo," then press Enter.

2. On the results page, click *More Options* near the top. You can limit your search right here by clicking an option like *Near you*, which will only show you tweets from people in your area. Or you can click *Advanced Search* at the bottom to set more limits, including dates and hashtags.

Advanced searches work a little differently from a mobile device. Open the Twitter app and tap the magnifying glass, then type your search terms. To refine your search, tap the icon in the upper-right corner of the screen.

Start an online scrapbook for creative projects

Get in touch with your creative side on Pinterest, the must-have website for DIY projects. Say you discover a craft or recipe you want to try, but you know you don't have time to do it now. You can save it for future reference by pinning it to your Pinterest account.

A "pin" is a visual bookmark — it holds that idea for you until you get around to tackling it. Be it a wreath for your front door, a makeover for that old dresser, a new way to wear your hair, slow cooker recipes, or party favors for a special celebration, Pinterest will help turn your dream into a reality.

Go to *www.pinterest.com* to create an account. Pinterest will help you choose topics to follow and set up your feed. You can save any article or image for later perusal by hovering your mouse pointer over it and clicking the red *Pin it* button. Then pin it to an existing board, if you have one, or create a new one. Boards help keep your pins neat and orderly. Don't see anything that strikes your fancy? Search for more things to pin in the *Search* box at the top.

Pin anything from the Web with just 1 click

You aren't limited to pinning only the things you find in Pinterest. With a single click, you can pin creative ideas from around the

Internet. All you need is the Pinterest browser button. To get it, visit *about.pinterest.com/en/browser-button* and click *Get our browser button*. Once installed, the Pinterest icon will appear in your Web browser's toolbar.

The next time you're on a Web page that you want to save for later, click the *Pinterest* button, name your pin, click *Pin it*, and assign it to a board. You can reopen that website any time you want by logging in to Pinterest, opening that board, and clicking the pin.

Discover this site's secret feature

Shhh — Pinterest has a secret to share with you! Thanks to the Secret button, you can create a board that no one else can see unless you invite them. It's a feature in your *Create a board* window. Click the button to the right of *Secret* to change it from *No* to *Yes*.

If you want to invite friends or family members to join your hush-hush project, type their names or email addresses in the *Collaborators* field.

You can also change the settings on an existing board to make it private. Simply open the board and click the *Edit board* button in the top-right side of the window. Now, only you and the people you invite will be able to see the items pinned there.

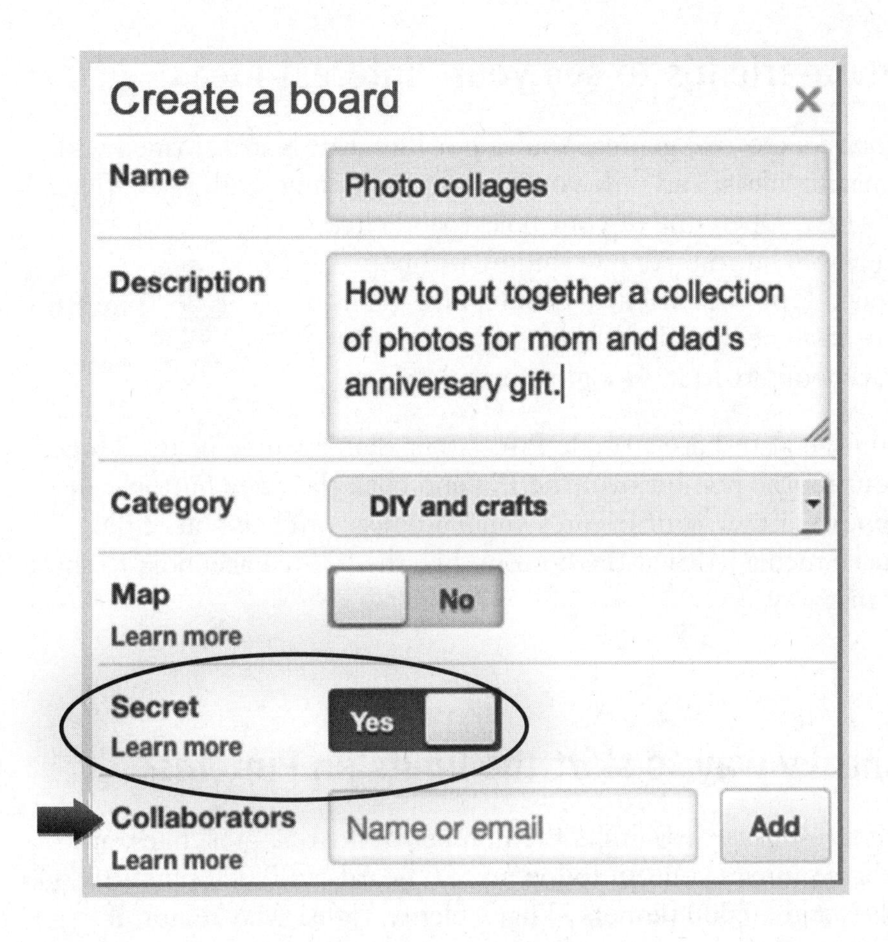

Update your Web browser for better pinning

For the best Pinterest experience, make sure you have the most recent version of Chrome, Firefox, or Safari on your computer. Plus, see if Javascript is enabled in your browser. Javascript helps Pinterest show you images and buttons. To check your browser's status, visit *Whatismybrowser.com*.

Invite friends to see your Pinterest ideas

You're a creative genius. You've got Pinterest boards loaded with amazing ideas, and now you want to share them with your friends. It's easy. Open one of your boards and click the *Invite* button. A new window will pop open. Type your friends' email addresses into the *Invite* field, or find friends who are on Facebook, Twitter, Google+, or Yahoo.

Invite

You can also share a single Pin, rather than a whole board. Hover your mouse pointer over the Pin and click the *Send* button that appears. Enter your friend's email address, or click one of the social media icons at the bottom, like the 'f' for Facebook, to share it that way.

Sneaky way to skirt the limits on Pinterest

Pinterest *generously* limits the number of boards, pins, likes, and fellow pinners you can follow to 500 boards, 200,000 pins, 100,000 likes, and 50,000 pinners. That's plenty, right? Maybe not, if you're a Pinterest addict. Here's the quick fix — create new accounts using other email addresses. Just make sure you keep up with your multiple accounts. Then pin to your heart's content.

Photos & videos

Focus on smartphone skills

4 easy ways to get photos off your phone

Cellphone cameras make it easy to snap a photo wherever you are, but do you really need 50 pictures of your beach vacation hogging all that memory on your phone? Get those photos off your phone, and you'll have plenty of room to take more. These options should work with tablets, too.

Connect to your computer. Plug your phone into your computer's USB port using the cable that came with the phone. The Photos app should open automatically. If not, find the program and launch it yourself.

On a Mac, press the *Import New* button. On a PC, the Import button looks like a box with a downward arrow inside. Click on it, check which photos you want to import, and click Continue > Import.

iPhones pair easily with both Windows PCs and Macs, but you'll need an app to move photos from an Android phone to a Mac. Android File Transfer is free (with a 4 gigabyte limit) and available at *www.android.com/filetransfer.*

Transfer using Wi-Fi or Bluetooth. If you want to upload your iPhone photos to your Mac — or share them with other Mac users — just use AirDrop. Select the ones you want to send, and tap the share box with the upward arrow in the bottom-left corner. Then tap the image of the nearby computer or person you want to share with.

For Android phones and PCs, you'll need a third-party app to perform wireless transfers. One of them, Filedrop, is even free.

Use a syncing cloud service. Dropbox, Google Drive, iCloud Photo Library, and Microsoft OneDrive are cloud services that provide the most hands-off ways to get photos off your phone. They can automatically back up photos as you snap them, and they let you access your photos from multiple devices. However, your free storage space is limited. Learn more about cloud-based storage options on page 261.

Back up your photos to an online service. These services don't sync to your hard drive, but you can download your images whenever you want. Android or iOS systems can access Flickr and Google Photos via an app or website. If the free space isn't enough, you'll need to pay for more.

Snap a selfie like a pro

You may think selfies are silly, but when there's no one around to take that great picture of you in your new car, you might have to bite the bullet and snap one. Don't run out and get a selfie stick, though. Just use these tricks to take the perfect smartphone selfie.

Turn on the right settings. On an Android phone, open the camera and tap the reverse camera icon, or tap Mode > Selfie. For a selfie with more than one person, tap Mode > Wide Selfie to fit everyone in the shot. On an iPhone, open the camera and tap the camera icon with a reverse symbol.

Pay attention to lighting. If you're taking your photo indoors, just use a simple light source. Stay out of direct sunlight outside to avoid squinting. Experiment with your flash but use it sparingly. On an iPhone, you can apply filters before you take a photo by tapping the circles in the bottom-right corner of the screen.

Change the exposure to focus on you. Your smartphone camera tends to autofocus on the brightest spot in the frame. If that's not you, then tap the spot on the screen that needs to be brighter on an iPhone. On an Android, look for a small box where the camera has focused, and drag it to the darker section to lighten it up.

Play with the automatic timer. Both Androids and iPhones have a timer that can delay taking a photo by up to 10 seconds. Just tap the timer icon after the front-facing camera is on. On the Android, it reads *Off* by default. The iPhone even shoots a burst of 10 photos when time is up, so you can pick the best shot.

On an Android phone, you can also turn on *Gesture control* under *Camera settings* (the gear-shaped icon) when you're in selfie mode. Once it's activated, you can simply show your palm to the camera, and it will take your photo two seconds later. On the iPhone, plug in the earbuds that came with the phone, step back, and use the volume-up button to snap a photo.

And don't forget the most important tip of all — shoot from above to avoid an unflattering double chin!

When you should never use flash on a phone

You grab a quick photo of friends at dinner, but the picture looks washed out. It's the curse of the automatic flash. Many cameras have their flash setting on Auto, so it fires whenever the sensor thinks the time is right.

Unfortunately, it's often wrong. Washed out colors, flat images, and dark backgrounds are just a few of the problems caused by automatic flash.

Your best bet is to keep your flash setting off and only use it if you take a photo in a dark room or outdoors at night. To change the settings:

- on an Android phone or tablet, tap the camera icon and look for a > symbol in the top-right corner. Tap > and then look for the flash lightning bolt icon. Your options are *Auto, On*, and *Off*. Tap the icon until it cycles to *Off*.
- on an iPhone or iPad, tap the camera icon and look for the lightning-bolt symbol in the upper-left corner of the screen. Tap it and select *Off*.

So what do you do when you want to take a picture in dim light? Check out these two solutions.

Adjust the exposure. You can change the ISO setting, which affects how sensitive your camera's sensor is to surrounding light. If you're in a dark environment, try changing the ISO to make the scene brighter.

To adjust the ISO on an Android phone, tap Mode > Pro > ISO, and slide the bar to a higher ISO number. On an iPhone, tap and hold the screen where you want to focus the image. A yellow box appears with a sun icon to its right. Slide the bar next to the sun up to add more light.

Supplement with other light. Turn on a nearby lamp, or ask a friend with a flashlight app on his phone to stand nearby and shine the light on your subject. If the light seems too harsh, drape a white napkin over it.

Perfect pictures without the shadows

Taking photos of people on a bright, sunny day? Your subjects will probably end up with squinty eyes and shadowy faces. That's when your flash will really come in handy.

Try to keep the sun off to the side of your subjects, and use the flash to "fill in" the shadows. Just open the camera app and tap the flash (lightning bolt) icon to turn it on.

You can also use the High Dynamic Range (HDR) option to balance the light and dark parts of your photos. The camera takes three pictures at different exposures, from very light to very dark, then layers them into one picture. It's great for portraits or landscapes.

- On an Android phone, look for *HDR* and choose *Auto, On*, or *Off*.
- On an Android tablet, tap Mode > HDR (Rich tone).
- On an iPhone or iPad, tap *HDR* to turn it on or off.

You may be tempted to use HDR all the time when you see how beautiful your photos look. Don't. Because it blends three photos into one, you'll be disappointed when shooting moving objects, brightly colored subjects, or high-contrast photos like silhouettes.

Grab great photos of subjects on the move

That perfect picture of two beautiful butterflies feeding on your snapdagons has turned into a colorful blur. Blame shutter lag — the half second it takes for the camera to focus, calculate the right exposure, and finally take the picture.

Next time, use your smartphone camera's auto-focus and auto-exposure lock. The AE/AF Lock is handy when you want to grab a photo of something that moves quickly, or when your camera auto-focuses on something in front of or behind your subject.

On an iPhone, hold your finger down on the spot where you want to focus your shot until you see the yellow square blink twice and you get an *AE/AF Lock* message on-screen. You can now shoot pictures immediately without waiting for the camera to refocus.

On an Android phone, you have two options for action shots. The easiest one is to open your camera's Settings and turn on *Tracking AF*. Then return to the camera and tap the subject you want to track. Tap the shutter button when you're ready to take a photo.

If the lighting is bad or the subject is far away, you can lock in the exposure and focus with the AE/AF setting. Tap your subject and hold it until the *AE/AF Lock* circle appears.

Never take a blurry picture again

Some people never experience the thrill of seeing a celebrity. And some do, but blow it by snapping a quick, blurry photo. Even if getting good selfies with celebs isn't your top priority, clear photos should be. Try out these simple tricks, and you'll never take a blurry picture with your phone again.

Stabilize the camera yourself. You may be part of the problem if you rush your shots. Before taking a picture, pause a few seconds and give the camera time to focus. If you're worried about your hand shaking, rest your elbow on a shelf, car fender, or any other stable object. You can also get an inexpensive tripod for your smartphone, which is especially useful if you want to be in the photo.

Get creative. If there's no surface around, hold your arm against your body, place your phone in your palm, and use an alternative way to click the shutter, such as the "up" volume button or voice control.

And don't forget about the camera's self-timer. It's not just for group shots and selfies. When you want to be absolutely sure you won't jiggle your phone, place it on a level surface and set the timer to take your picture.

Use the built-in setting. Some of the latest smartphone cameras, including the Samsung Galaxy S6 and the iPhone 6 Plus, have built-in optical image stabilization (OIS). This means the camera's sensors adjust and stabilize the image before you even take the picture.

That's a big improvement over the digital stabilization used on older phones, which sharpens up a blurred photo after it's been taken. If you're lucky enough to have OIS, make sure the feature is turned on.

Use Burst Mode for fast-moving photos

What's almost as cute as a dog and kids playing together in a backyard? The photos you take of their rough-and-tumble antics, of course!

Snapping a clear picture of kids and dogs in action is almost impossible, though, without a handy feature called Burst Mode. Instead of pressing and releasing the shutter button to take a single photo, you can hold it down to take several fast shots in a row.

On an Android phone. Point at your subject, then press and hold the shutter button. The camera will keep snapping photos for as long as you press the shutter, up to 30 photos. A *Burst* icon will tally your pictures. Tap it when you finish, then tap the *Burst* icon again to scroll through your pics. When you see one that you like, choose *Save as New* from the menu at the top.

Want to delete the rest? Tap the *Back* button to return to the beginning of your Burst album, then tap the *Trash Can* icon at the bottom of your screen.

On an iPhone. Just hold the shutter button down and your camera will snap up to 100 photos, plus keep a running count. Tap the *Burst* icon to page through the shots, tap *Select*, and choose which ones to keep by tapping their checkboxes. When you're finished, choose *Keep Only Favorites* to delete the rest, or *Keep Everything* to keep all of them.

Burst mode can't do all the work for you, of course. Set yourself up for success with a few quick tips to improve your action shots.

- Avoid taking photos in front of signs, poles, and other distracting backgrounds.
- Get as close as you can to your subject.

- Keep the sun at your back so that your subjects' faces are clear and well-lit.

Panorama Mode fits everyone in the photo

Big family reunions can feel like one long flurry of photo-taking. That includes group photos. And how on earth do you fit everybody into one frame?

Take a panorama shot — a wide-angle photo that's not limited to the size of your screen. Nope, you don't need a fancy $1,000 camera. Chances are, you can do it on your smartphone!

Most smartphones have a panorama feature. They take a series of images, seamlessly stitch them together, and create one wide photo that captures 240 degrees of family fun.

On an Android phone or tablet. Tap the Camera icon to open the Camera app, then tap the word *MODE* and choose *Panorama*. To start the photo, just tap the shutter button. Slowly sweep your phone in one direction. Be sure to keep your subjects pictured inside the translucent rectangle and within the center strip.

Your device will tell you if you need to go slower or if you stray too far up or down. Tap the shutter button again to end your photo.

On an iPhone or iPad. Tap the Camera icon, scroll through the *Photo* options on the right, and choose *PANO* mode. Then follow the same steps as for Android devices.

Panoramas are also great for taking photos of beautiful scenery. Plus you can take them vertically to capture the full majesty of tall buildings, trees, or monuments. Keep in mind that the image file is bigger, so it will take up more storage space on your gadget.

Make must-see videos with your smartphone

Gone are the days when your cousin spent every family holiday with a camcorder on his shoulder. Those videos could be painfully dull, and the shaky camera and random zooming didn't help, either. Now you can take over documentary duties, because you can make fantastic videos with your smartphone.

Go sideways. Turn your phone horizontally before you start recording, unless you're filming something exceptionally high or tall. If you hold your phone vertically while recording, you'll see black bars on each side of the video when you play it back on your computer.

Get stable. "Shaky cam" might be popular in films, but not in smartphone videos. Since you'll be holding your phone for longer than a second or two, hold it close to your face with two hands, and rest your elbows on your chest or stomach for added steadiness.

Follow the action — the pros call it "panning" — by slowing turning your torso, not your arms. For shoots lasting longer than a minute or two, consider investing in a monopod or tripod.

Move closer. Instead of zooming in with your phone, which can make images fuzzy, get as close as you can to the action. And never zoom and pan at the same time. It can be disorienting for your viewer.

If you really enjoy making phone videos, consider buying a lens made specifically for smartphone cameras. You can find microphones made just for smartphone recording, too.

Look elsewhere in this chapter for tips on taking photos with your camera — many of the same tricks apply to video.

Edit photos on your phone, without costly software

Forget the bad snapshots you've taken over the years — you know, the ones you thought would turn out great until you picked up your film at the drugstore. You'll never again face that unpleasant surprise of blurry, crooked, or otherwise embarrassing shots.

Not only can you immediately delete photos on your phone that don't turn out well, but you can turn mediocre ones into pro-grade shots by editing them right on your smartphone.

On an Android phone. You can set up special effects before you even snap the picture. Open your Camera app and tap the *Effect* icon to see all the ways you can change the look of your shot. Tap one, then snap your photo.

Or tap Mode > Pro to play with individual settings like light sensitivity and color tone. Love taking photos of your meals or want to erase the wrinkles in your selfie? Tap *Download* in the main *Mode* menu and search for *Food, Beauty Face*, and more tools, many of which are free.

After you take your picture, you're just getting started! Tap that photo to open it, then tap *Edit* to see tons of options.

- Automatically adjust colors and shadows with *Auto adjust.*

- *Rotate* or *crop* your photo to get the best composition.
- If you took multiple shots, tap *Collage* and get creative.

Under the *Edit* menu, you can also tap *Photo Editor* for yet more options.

- Fix crooked backgrounds or skewed horizons with *Adjustment*.
- Change brightness or saturation with *Tone*.
- Choose *Effect* to apply filters after you take the photo.
- Tap *Portrait* to fix red eye and make other cosmetic changes.

The Photo Editor on an Android tablet has fewer features, but you can still rotate, crop, adjust colors, or apply filters to the pictures you take with it.

On an iPhone. Snap your picture, tap the tiny image of your photo in the bottom corner, then tap *Edit*. You can auto-adjust the image by tapping the wand icon; or tap one of the other icons to rotate the picture, apply a filter, or play with color saturation.

If you installed a third-party, photo-editing app from the App Store, you can access it by tapping the ellipses.

Preview versus Photos: which Apple app works best?

Apple gives you two ways to view and edit photos on your Mac — the Preview and the Photos apps. But why do you need both? Because these apps have different functions.

With Preview, you can do basic photo editing. With Photos, you can do more in-depth editing plus organize your photos into Albums and Collections for sharing. A scaled-down version of Photos is also available on the iPhone and iPad.

Try basic touchups in Preview. All photos on a Mac will open in Preview unless you tell the computer otherwise. So open your image in Preview and click:

- *File* in the Menu bar along the top to rename, print, or duplicate an image.
- Tools > Adjust Color to open a box where you can play with saturation, tints, and more. The same Tools menu also features *Adjust Size*, where you can change the resolution or dimensions of the photo.
- the toolbox-shaped *Markup* icon in the toolbar. Use the buttons that appear to draw on, crop, or otherwise alter your picture.

Do deeper editing with Photos. Open the Photos app, then click File > Import to find and open your picture. After the app imports your photo, double-click the image to open it, and click the *Edit* button in the upper-right corner.

Editing tools like *Rotate* and *Crop* appear on the right, some of the same ones you get in Preview. What Preview doesn't offer, however, is the ability to try nine different filters, retouch your photo to erase blemishes, or let Apple improve your image automatically by clicking *Enhance*.

Adjust is the most involved feature. Clicking that button allows you to tweak three different settings independently of each other — *Light, Color*, and *Black & White*.

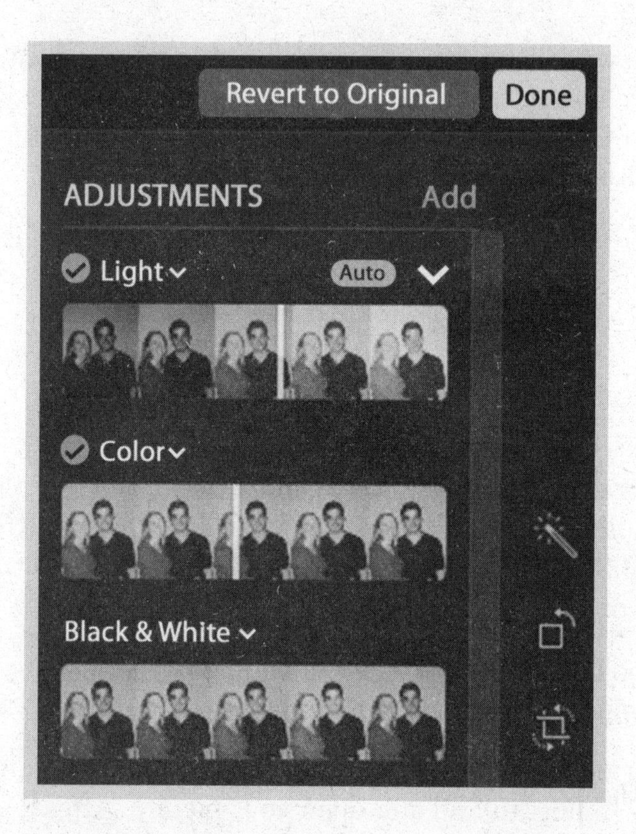

Each setting features a mini-filmstrip of your photo, with sliders that you can drag right or left. As you do, your photo will change, too. Point at each setting, then click on the drop-down arrow that appears to the right to see submenus for each option.

Click *Add*, to the right of *Adjustments*, and you'll get a drop-down menu with even more settings you can change, from sharpening the image to reducing noise.

Press the M key on your keyboard to flip between the original image and your edited version. Hate what you've done? Don't worry, you can click *Revert to Original* any time. When you're happy with your photo, click *Done* to save your changes.

Punch up your pictures with Windows Photos

Think you need professional photo-editing software to fix crooked shots, unwanted backgrounds, or red eye? Think again. The Windows 10 Photos app lets you easily fix all of those issues and more — plus organize and share your pictures.

Open the app. Click ■ to open the Start Menu, then click the *Photos* tile. The app will grab any photos in your Pictures folder and in your OneDrive account. You can block it from accessing OneDrive by clicking *Settings* in the bottom-left corner of the window. Then click the button beneath *Show my cloud-only content from OneDrive* to turn it off.

Improve photos automatically. Click a photo to open it and notice the ⁂ that appears above it. The Enhance icon looks like a magic wand because it works wonders. Click this tool to straighten a crooked background, make red eye disappear, or tweak the contrast and brightness. It doesn't alter your original photo, only how you see it on-screen.

Take control with editing tools. Want to spiff things up yourself? Open your editing tools by clicking the pencil-shaped icon. A smorgasborg of options will appear down the right and left sides of the photo. The tools are simple to use. Some present you with a dial you can drag in a circle to

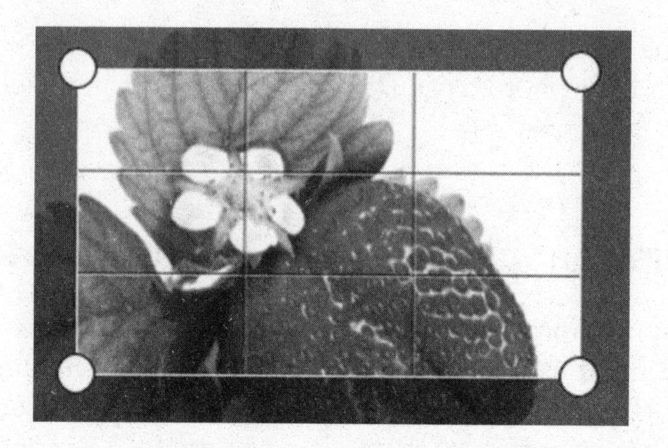

crank an effect up or down. Others give you a box to drag around and position over the photo.

- Click *Retouch* to erase blemishes and wrinkles.
- Make a certain color pop by clicking *Color Burst.*
- Click *Basic Fixes* to see tools like *Rotate, Crop, Straighten*, or *Red Eye.*
- Change the whole look of your photo with one of six *Filters.*
- Click Effects > Selective focus to place a circle or oval over part of the image and blur out the rest; or soften and darken the edges of your photo with Effects > Vignette.

Share finished photos. Click to email your photos to a friend, upload them to an online storage service like Dropbox, or post them on Facebook or Twitter.

Organize your albums. Photos automatically groups your pictures into albums based on things like date and subject matter, but you can make your own albums, too. Return to the main *Photos* window and click *Albums* on the left. Then click the plus sign (+) in the upper-right corner of the window.

Click each photo you want to add to your new album. When you're done, click the check mark (√) at the top. Name your album and click 🖫 in the top-right corner.

Find your photos faster by tagging them

Computers are supposed to make life easier, but your digital photos are just as messy as the ones you used to store in a shoe box. Most of us just aren't good at keeping them organized. That's why you should tag them with keywords as soon as you put them on your computer. Then you can use those keywords later to find the pictures you want, fast.

In Windows. Import photos from your camera or smartphone and save them in your Pictures folder. Then click ⊞ > File Explorer > Pictures, and open the folder that houses the photos you want to tag. Make sure you can see the *Details pane* on the right side of the window. If you can't, click the *View* tab in the top-left corner and then click *Details pane.*

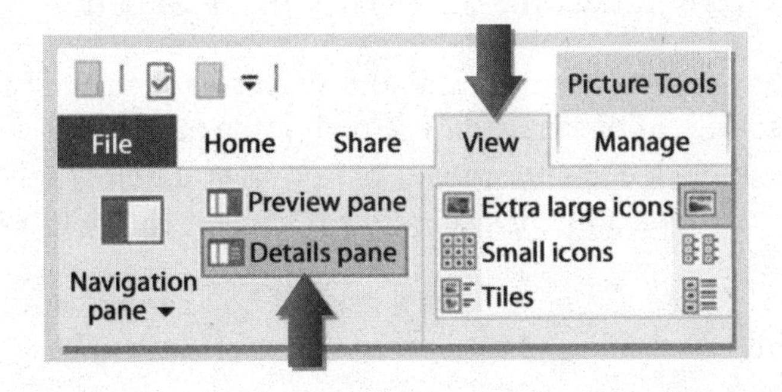

Select the photos you want to tag, and click *Add a tag* on the right side of the window in the *Details pane*. Type whatever keywords will help you find these pictures later. Try words like "Family," "Vacation," or "Holidays," or get specific with descriptions like "Grand Canyon" or "Seattle Zoo." You can assign more than one tag — just separate them with a semicolon, such as "Vacation; Grand Canyon; June 2016." Then click the *Save* button.

You can tag photos one at a time, but if you want to assign the same tag to lots of pictures, do them all at once. Hold down the Ctrl key and click each photo you plan to tag. Then type your tags in the *Details pane* and click the *Save* button at the bottom.

On a Mac. Apple is a neat-freak's dream. You can tag any type of folder or file (including photos) with colorful circles, keywords, or both. While transferring the photos to your computer, type a tag into the *Save As* field.

For images already on your computer, press Control while clicking the file, then click *Tags* in the shortcut menu that opens. Choose an existing tag or type a new one, and press Enter after each. You can also click a colored dot. It will show up next to the file name in your *Finder* window.

Need to find a group of photos? Now it's a breeze. Type a tag word like "Family" into the *Search* box in Finder (Mac) or File Explorer (Windows), and press Return. Your photos will come up right away.

Or let Cortana, the Windows 10 virtual assistant, find them for you. All you have to do is say, "Hey Cortana, show me my photos from ..." and name a month, year, or keyword.

Safely store your photos on an external hard drive

You thought your computer was a great place to store photos — at least until they started filling up your hard drive and slowing down your computer. The solution? Move your photos onto an external hard drive and delete them from your computer.

You'll free up storage space plus protect your precious mementos if your computer dies one day. Buy the biggest external hard drive (HD) you can afford, then plug it into your computer to start moving your photos.

On a Mac. Open the Photos app. In the Menu bar at the top of your screen, click Photos > Preferences > General. Click the *Show in Finder* button, and a Finder window will open, highlighting the file that contains your Photos library. Your external HD should also appear on the left side of the Finder window under *Devices.*

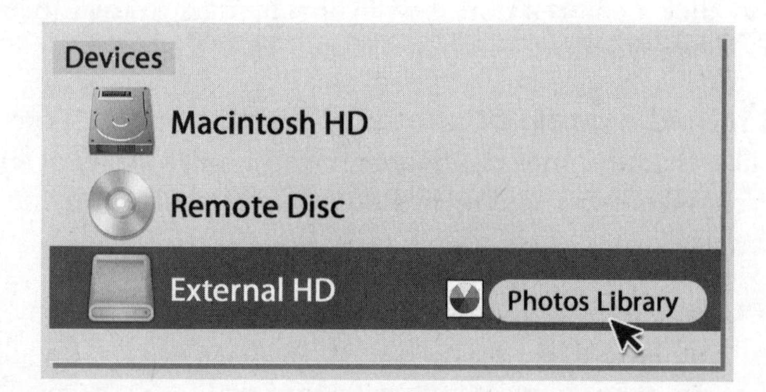

All you need to do is move the Photos Library file to your external HD. Drag the *Photo Library* file to the icon for your external HD. Now tell the Photos app to look for your pictures on the external HD, not your computer, when you open them.

1. Close the Photos app, then press *Option* while opening the app again. This opens a window titled *Choose Library*. Click *Other Library*.

2. In the mini-Finder window that opens, find the Photo Library file on your external HD, and then click *Open*.

On a PC. Click the ■ on the left end of the Toolbar, then click *File Explorer*. Your external HD will appear on the left side of the *File Explorer* window. Drag the *Pictures* icon to the external HD to move your photos.

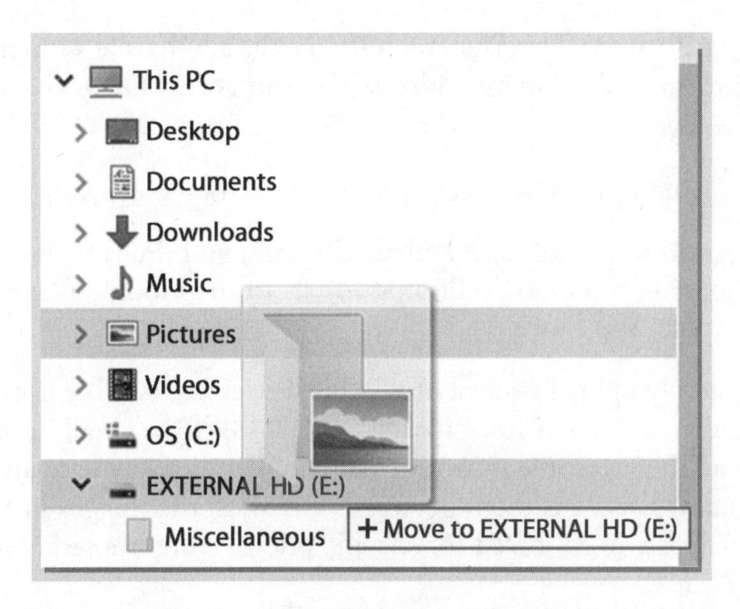

Make sure that the Photos app knows where to find your pictures from now on. Just open Photos and click Settings > Add a Folder, then choose the external hard drive.

Cloud storage: a safe place for your precious photos

Photos can quickly pile up unseen. Then if you lose your phone or your computer crashes, those priceless images are gone forever. You need an online, cloud-based storage service to automatically save your photos. Plus, no house fire can touch them if you store them online. Here are three options that keep things synced so you can access your images anytime.

Google Photos. This is a cinch if you already have a Google account, such as a Gmail address. You can access Google Photos from:

- any computer by logging in to your Google account online.
- Android phones and tablets, iPhones, and iPads by downloading the Google Photos app from the Google Play or Apple App Store.

Next, simply upload your photos. This service has two big upsides — 15 gigabytes (GB) of free storage (or 100 GB for a small, monthly fee), and the awesome power of Google search. Google analyzes your photos, assigns them keywords based on the images in them, and enables you to search for specific photos using those keywords.

For example, it might group pictures of the family dog in a "dogs" collection. Google can even identify faces and ask you to assign names to people in those photos. Plus you can search photos by year or location. The only downside? It compresses large photos, which can degrade their quality.

Apple Photos and iCloud Photo Library. The Apple Photos app can only be used on Apple devices such as iPhones, iPods, iPads, and Mac computers. It has some of the same bells and whistles as Google Photos, including facial recognition and the ability to search by date, year, and subject.

Combine it with Apple's iCloud Photo Library, an online storage service, and you can easily store your images on the Internet. The iCloud Photo Library gives you 5 GB of storage for free. If you need more space, you'll have to pay.

Flickr. When it comes to storage space, Flickr wins. You get 1,000 GB for free, enough for about 650,000 full-resolution photos. All you need is a Yahoo! account. Then you can upload your photos directly to the website *flickr.com*. Or get the Uploadr tool at *flickr.com/tools* and install it on your computer.

Flickr makes a similar tool for smartphones and tablets, too, called Auto-Uploadr, that will automatically save every photo you snap to the Flickr website. Flickr tags your pictures with keywords, but you can add your own as well. This site lets you easily share photos on social media. Just send people a link to a photo or album, or designate which people can see your photos.

Share your pictures with the world on Instagram

Instagram is the place to be for photographers who enjoy social media. On this mobile app, anything goes, from the swirl of steam rising artistically from your morning coffee to the silliness of your dog's new sweater. Instagram gives you a quick and easy way to share a bit of your day.

Snap and share your first photo. To start "Instagramming," download the app from the Google Play or Apple App Store onto your phone or tablet. Tap the app's icon and sign up using your Facebook account or email address. Tap the silhouette to add a profile photo, if you like, and you're ready to upload and share your first picture. Tap the ◙ in the bottom-center of your screen, and you will have two options.

- Take a new photo. Tap *Photo*, then tap the large circle to take a picture with your regular camera; or tap ↻ to snap a "selfie" with the front-facing camera. Turn on the flash, if needed, by tapping ⚡.
- Share an existing photo. Tap *Gallery* (Android devices) or *Library* (Apple devices) and choose any of the photos saved on your gadget.

Get ready for fun with Edit. Tap the → in the top-right corner of the Instagram screen to edit your image before sharing it. Time to experiment!

- Instagram is known for its filters. Tap one of the filters along the bottom of the screen to instantly change the look of your picture.
- Tap ◎ to change lots of other properties, including the brightness, contrast, or colors.

Once you're done, tap → or *Next* to share it. Give your photo a brief caption. Hashtags — phrases that start with # — are a big deal on Instagram. People use them to find photos with similar subjects.

If you post a photo of your dog, for instance, you could tag it by typing a phrase such as #dogsofinstagram or #dogsinclothes in the caption. Ready to share it with the world? Tap the check mark (√) or *Share*.

Wow your friends with amazing Instagram photos

Get ready to take your pictures from OK to amazing. Your Instagram followers will rave when they get a peek at your pro-quality shots.

Post in portrait or landscape. Not a fan of Instagram's squared-off style? You can share more traditional portrait- or landscape-style photos, too. Open Instagram and tap ◙, then tap *Gallery* or *Library*. Tap a picture you want to share and tap ⤡ below the photo to change its orientation.

Create collages in Layout. Combine several separate photos into one image with ease. Open your Gallery or Library and tap ⊞ to download the Layout app. Now, when you want to post a photo collage, open your Instagram Gallery/Library and tap . You can then scroll through your pictures, tapping up to nine to include in your collage.

Or tap *Photo Booth*. Your phone will snap four photos of you with the front-facing camera and stitch them together in a variety of collages. Just pick the version you like best. Tap your collage to edit the photos in it. You can even rearrange the pictures by dragging them around and dropping them into place.

Take artsy shots with ease. The Tilt Shift feature focuses on one area of the image and makes the rest appear soft and blurry. Depending on how you use it, you can make a photograph of a real city look like a miniature model, or put the subject of your photo in sharp focus while leaving the background blurry.

Snap your photo and tap . Swipe through the options at the bottom until you find *Tilt Shift*.

Rescue a bad snapshot. Lux can help save photos that are underexposed or too dark. Tap on the *Edit* screen. Slide the circle along the line that appears to brighten your picture or give it more contrast. Lux works especially well on photos of buildings or landscapes.

Instagram is packed with built-in features, but don't forget the many third-party apps you can download and add on to it. Cymera and VSCO, for example, give you the power to change all kinds of settings, as well as filter to your heart's content.

To overlay text, frames, stickers, or geometric shapes onto photos, try A Beautiful Mess or Aviary. Get them through the Apple App Store or Google Play.

Shrink photos to make them easy to email

Attach a high-quality photo to an email, and there's a good chance it will bounce back as undeliverable. That's because email programs can't send huge files. And the higher a photo's resolution, the bigger the file.

Resolution refers to the number of dots per square inch that make up the photo — not the photo's actual size in inches. High-res photos take up more memory, but they look better when printed at large sizes because they're sharper and more detailed.

That said, you don't need to send your friends a high-resolution photo if they're going to view it on a computer or phone screen. A small-resolution image, such as 640 x 480 or less, will work just fine — and it won't bounce back if you email it.

Most smartphones will turn high-res photos into low-res ones when you email or post them directly from your phone. If you're sharing the photos from your computer, though, you'll need to shrink them yourself.

- Macs make it easy. Open the image with your Preview app. Then click Tools > Adjust Size in the Menu bar to open the *Image Dimensions* box. Click the drop-down list beside *Fit into* and choose from several standard resolutions. Pick 320 x 240, for instance, and your picture will become a much more manageable size. Click *OK*, then attach the photo to your email and send.

- Microsoft makes it hard to do the same thing in Windows 10. Your best bet is to shrink or compress photos using your email program, or to share them through a cloud service such as OneDrive or GoogleDrive. Learn about emailing photos and other large files on page 138, and about cloud services starting on page 261.

Movies & TV

New ways to watch

How to find the videos you need, fast!

YouTube is the how-to capital of the world. With millions of videos detailing everything from how to tie a tie to how to get a six pack in three minutes, you don't have to look any further than your computer screen to find a solution to almost anything. But to find the videos you need, fast, it helps to have a few tricks up your sleeve.

Filter your finds to the types of videos you want. Rumor has it, more than 300 hours of video content is uploaded to YouTube every hour. That's a lot of videos to choose from. Luckily, YouTube provides an advanced search option in the Filters dropdown menu.

First, type what you're looking for in the search box and press enter or return. At the top of your search results you'll see a little box with the word Filters. Click it to search by type, length, features, and more.

Filters ▾

Upload date	Type	Duration	Features	Sort by
Last hour	Video	Short (< 4 minutes)	4K	Relevance
Today	Channel	Long (> 20 minutes)	HD	Upload date
This week	Playlist		Subtitles/CC	View count
This month	Movie		Creative Commons	Rating
This year	Show		3D	
			Live	
			Purchased	
			360°	

This can come in handy with almost any search. Say you want to look up how to fix a problem you're having with your phone. Technology changes faster than you can say, "I just got the latest model," so those old videos aren't going to help. Filter results by upload date to get only the most recent videos.

Limit your search to words in the title. Suppose you watched a video a few days back called "How to troubleshoot a computer," and you want to find it again. If you type those terms into the search box, you'll get results with any of the words or synonyms. But if you type "allintitle: how to troubleshoot a computer," you'll only see results that contain those exact words in the title. It's a simple step with big benefits — more watching and less searching.

Master videos without messing with your mouse

Fumbling with your mouse to pause, play, or adjust a video is not your idea of a good time. You're not the only one who feels that way. So you won't be the only one delighted to learn YouTube has secret shortcuts to help you avoid using your mouse. Some of these shortcuts only work if your pointer is within the video frame. Select the video by clicking somewhere on the video player before using the keys.

Key	Action
Space	Play/Pause
k	Play/Pause
j	Jump back 10 seconds
l	Jump forward 10 seconds
m	Mute
f	Full screen
Esc	Exit full screen
Right arrow	Fast-forward
Left arrow	Rewind
Shift+>	Increase speed
Shift+<	Decrease speed
Home	Jump to beginning
End	Jump to end
Numbers	Jump to section (For example, 0 jumps to beginning and 5 jumps halfway.)

Spend 'quality' time with sharper, larger YouTube videos

Once upon a time, YouTube only had one type of video available — standard quality. But no more. These days, this video-sharing empire gives you multiple ways to stream videos. You just have to know how to adjust your settings.

Spend some quality time with YouTube. Video quality is automatically set to Auto, which uses the highest quality based on video player size. But you can manually change the quality settings — if your slow connection makes it difficult to download HD videos, for instance. Click the settings gear, then select *Quality*, and choose the quality that works for you.

You can change your quality settings permanently by signing in and clicking your account icon in the top-right corner. Next, select the gear icon > Playback > Always choose the best quality for my connection and player size.

Don't know how to sign in? If you have a Google account, you already have a YouTube account. Just use the same login credentials.

Size up your video to improve your viewing experience. The size of your video player automatically adapts to the size of your computer screen, but if you want to make the video bigger, YouTube gives you a few ways to do that.

- Take up the whole screen by clicking the Full screen icon.
- Make the video larger without going full screen by clicking the Theater mode icon.
- On mobile devices, switch to full screen by tapping the screen, then the Full screen icon.

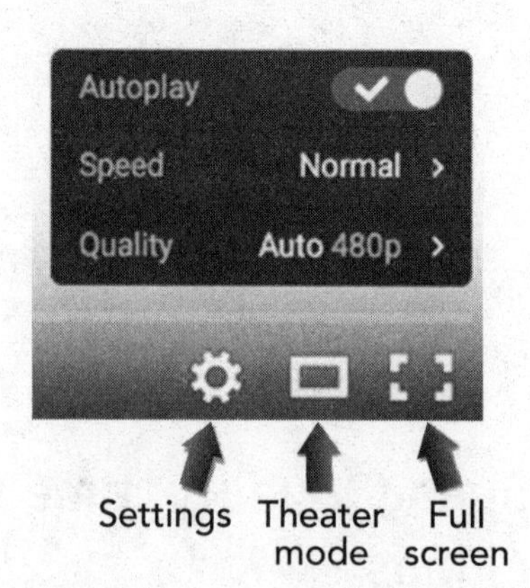

Start that YouTube video at that exact 'wow' moment

Have you ever waded through an entire YouTube video for that one special moment when "it" happens? Like that time your friend sent you that video URL with the explanation, "The best part is in the middle!" Wouldn't it be nice if the video URL opened straight to the part she was talking about. Well, now you can share a URL that pinpoints that "it" moment and ignores the rest!

Just pause the video at the time you want to capture (right before the "it" moment). Right-click the video player, and select *Copy video URL at current time*. Paste the URL wherever you like. When you click the URL, it will start the video right where you want it to.

Cut your cable and switch to easy streaming

Cord cutting — it's all the rage. No, people aren't literally taking the scissors to their electrical cords. This cord cutting is about severing ties with your cable company. Cable costs keep going up, and lousy service has created many unsatisfied customers. But switch to a streaming plan, and you could be on your way to cutting your bill in half. No wonder it's so trendy.

With online streaming services, you can stream what you want to watch, when you want to watch it by viewing videos over the Internet instead of paying for a cable package. Simply pick a service based on your preferences.

The most common ones include Netflix, Hulu Plus, and Amazon Prime Video. They're perfect for streaming TV shows, movies, and original series. Services like PlayStation Vue and Sling TV even allow you to stream live TV. Each service has different content available, so check them out to make sure you'll have access to your favorites.

Of course, streaming isn't for everyone. If you still want to watch a lot of cable shows, especially sports, you will end up purchasing additional streaming services, which can add up.

But if you only want some cable channels, a hybrid cable-plus-streaming approach could be your best option. It's flexible enough to fit your budget and eliminates those channels you never watch. Just pay for the cable-and-broadband subscription you want, and supplement it with streaming services.

You can get to most of the major services on the Web, but to watch on the big screen, you'll need a TV streaming device called a set-top box. The most popular are Roku, Amazon Fire TV, Apple TV, and Google Chromecast, which cost from $35 to $100. Similarly, you'll want to look into these to make sure they have access to your preferred streaming service.

Stream online shows and movies for free

Internet streaming services like Netflix give you access to a wide range of media, but before you pile on subscriptions, take a quick peek at the free services. You just might find everything you want to watch — for the price of zilch.

Popular free sites for watching movies and more include Hulu, Crackle, Popcornflix, and Snagfilms. They don't have as much variety as the paid services, but if you're looking for a quick fix, they just might do the trick.

And don't forget about network websites like *abc.go.com*, *nbc.com*, *cwtv.com*, and *cbs.com*. You probably won't find whole seasons, but if you miss an episode, you might be able to find it on the site within a few days. Some sites even allow you to watch live, depending on your location.

Not all of these sites have apps that will stream on your TV, but hey, you can't argue with free.

Simple antenna can kick back the cost of cable

You hear the word "antenna," and immediately you think of the long rabbit ears people used to position strategically around the house. But antennas have received a major upgrade since those days, and now people are kicking cable to the curb and switching to the much-cheaper antenna.

How much cheaper? Well, an antenna could cost $30 to $100. That's a one-time fee, and then all your channels are free. You'll be amazed at what you can pick up on today's small but high-powered antennas. You get most of the big broadcast channels, such as ABC, NBC, CBS, Fox, CW, and PBS, plus independent channels.

While your location determines if you're a good candidate for an antenna, you have to consider if the loss of certain broadcast networks like ESPN is worth the switch. Check out options for your location at *tvfool.com*.

Record your favorite shows — no cable necessary

So you finally cut the cord. And now you don't know how you're going to record your favorite shows. Relax, you can still record a show even if you don't have cable. You just need to get a digital video recorder (DVR) that connects to your antenna.

Crowd favorites like TiVo Roamio, Tablo, and Channel Master DVR+ let you record live shows from over the air. DVRs can get pricey, especially if you pay a subscription fee to access channel information, so make sure you read all the fine print before you purchase.

Stream HD videos from small screen to big screen

Crowding around a teeny-weeny screen is the quickest way to ruin a movie. But why watch videos on your computer or tablet screen when you could be watching them on your nice, big TV? That's right, Apple gives you two easy ways to share your screen with the whole family.

- On your mobile device or iPad, swipe up from the bottom of the screen to open Control Center and tap *AirPlay*. Then tap the name of the device you want to stream content to — in this case, your Apple TV. Next, turn on *Mirroring*.
- On your Mac, click the AirPlay icon ( ) in the menu bar, then choose your Apple TV under *AirPlay to*. If you don't see the option *Show mirroring options in the menu bar when available* under > System Preferences > Displays, your Mac doesn't meet the requirements.

If you have an older Apple device that doesn't support AirPlay, you can purchase an adapter from the Apple store. Apple Digital AV Adapters can be used with older and newer iOS devices to connect to a compatible TV, projector, or other external display. Like AirPlay, it will allow you to mirror your screens.

Protect your viewing habits from nosy advertisers

Are advertisers watching you watch TV? They can with the technology in new smart TVs. It's called automatic content recognition (ACR), and unless you disable this function, marketers could have a window into your viewing habits.

ACR is built into many smart TVs, including popular brands such as LG, Samsung, and Vizio. You can either turn it off during setup or explore your settings later on. The key is taking a moment to dig through those pesky privacy policies so you don't unknowingly agree to letting your TV track what you watch.

The 'ultra' cool way to save your favorite flicks — forever

Don't you hate it when you buy a video but can only watch it from home? Well, the era of DVDs may be coming to an end. A new kind of movie watching has come to town, and everyone is giving it a warm Hollywood welcome. It's called UltraViolet. And though it's ultra modern, the concept is pretty simple.

When you buy Blu-ray Discs and DVDs with the UltraViolet logo on the box, you also get a redemption code for a digital copy. Or you can eliminate the hard copy entirely and just buy the code online. Then, use the code to add the movie or show to your UltraViolet account at *myuv.com*. UltraViolet verifies and stores your movie and TV collection in the cloud, so you can watch your movies on any device, anywhere, and save them forever.

To watch your registered movies and shows on your favorite device, you need to use a service that works with UltraViolet. Popular services like Flixster, Vudu, and CinemaNow link to your UltraViolet account so you can stream and download your collection to your devices.

ULTRAVIOLET

The service you choose is up to you. You may end up using more than one because each service is not yet supported on all devices. But managing several accounts is not as confusing as it sounds because you can link your accounts together and use just one password.

You're probably wondering about the hoards of DVDs you already have. Don't worry, you don't have to buy new ones. To get a digital version of DVDs you've already purchased, you can use disc-to-digital services. Vudu, for instance, has partnered with Walmart to make this process more convenient for those who are technically challenged.

The downside is UltraViolet is not available for all media purchases. Some studios and distributors, including Disney, don't currently support UltraViolet. That means some of your DVDs will have to wait.

Music

High-tech tips for online tunes

7 free ways to stream your favorite music

Jam to your favorite music without purchasing a thing — sounds too good to be true. But it's the new standard in the music world, and these seven streaming services are at the movement's forefront. While you can upgrade most of these services to get rid of ads and access more features, they're all available for free and on the go — music to your ears.

Service	Features	Upgrade
Spotify	Spotify.com offers stations and playlists based on genres, trends, and even how you're feeling. So if your mood is best described as "Running Thru a Field of Smiles" or "Walk In Like You Own the Place," just a couple of Spotify's playlists, the streaming service has got you covered. Spotify is extremely social and lets you connect with friends through Facebook, so you can see what they're listening to and share your favorite tracks.	Premium $9.99/ month

Service	Features	Upgrade
Pandora	Pandora gets a lot of attention for its easy-to-use features. Go to *pandora.com* and type in your favorite artist, genre, or song, and Pandora creates a personalized station. Pandora also allows you to shuffle all or just a few of your stations, so if you're in the mood for variety, you can mix it up.	Pandora One $4.99/month
Google Play Music	Go to *www.play.google.com/music/listen#/now* and browse by genre, mood, activity, decade, and more. Then upload your own music collection. And if you already have a Google account, you don't have to sign up for anything new. Plus, there's one more perk. If you sign up for the paid subscription, you get YouTube Red for free, which allows you to watch ad-free videos.	Subscription $9.99/month
Slacker Radio	Slacker.com has the usual streaming features, but it has one thing other services don't have — a feature called Artist Takeover. It's where artists, actors, YouTube celebrities, and more take over as DJs.	Premium $3.99/month or Plus $9.99/month
iTunes Radio	Sign in to iTunes with your Apple ID, and click the *Radio* tab to get instant access to Beats 1, Apple's very own radio station. You can listen to the music and artists you love for free with limited interruptions.	Apple Music $9.99/month single $14.99/month family
iTunes Internet Radio	Did you know iTunes gives you access to live radio stations around the world? You don't even have to sign in with your Apple ID. All you need is the iTunes app. Go to View > More > Internet Radio to find your new favorite broadcast. Then, click the station name and drag it to the left to add a shortcut to your playlist tab.	—

Service	Features	Upgrade
iHeartRadio	iHeartRadio lets you listen to live stations from across the country at *iheart.com*, even without signing in. You can also create an account for personalized, commercial-free custom stations. iHeartRadio has an app specially designed for cars, trucks, and SUVs called iHeartAuto.	—

Listen anywhere with paid streaming

Free music streaming accounts are great, but they do have their limits. If you're tired of being interrupted by ads and having a limit on how many songs you can skip each hour, consider upgrading. The following two services don't have free versions. But they offer free trial periods, so you can see if they're right for you before you commit.

Sign up for a free 30-day trial with Amazon Prime Music. With Prime, you can listen to over a million songs — ad free. You also get access to personalized stations and playlists created by experts for every genre or mood. Music is stored in the cloud, so you can listen anywhere or download songs to listen offline. Some services limit the number of times you can skip a song, but for $99 a year, you can skip all you want.

If you're already signed up for any of the following Amazon memberships, you have access to Prime Music.

- Annual Prime membership

- Annual Amazon Student or Mom memberships
- 30-day Amazon Prime or Mom free trials

Try out Apple Music's free 3-month trial. If you eat, sleep, and breathe music, you may be happier with Apple Music. This service offers even more songs — over 30 million.

Like Prime, Apple Music has playlists and stations created by experts. Plus, you can download Apple Music content to your library, then listen offline with unlimited skips. Go to *www.apple.com/music* to begin your membership. Current pricing is $9.99 a month for single plans and $14.99 a month for up to six people on a family plan.

Reveal the iTunes menu bar on your PC

The first time you open iTunes in Windows 10, you may get a surprise — the Menu bar is off by default. That means *File, Edit, View*, and all those options you've come to rely on are hiding in plain sight. But where are they? Here are two simple ways to pull up that menu.

Access for a little while. If you just want to pull up the Menu bar temporarily, press the Alt key. Make it disappear by pressing the key again. Next time you open iTunes, the menu will be hidden once more.

Access forever. If you feel lost without the Menu bar, you can make it stay permanently by clicking the menu icon in the upper left corner of iTunes. Select *Show Menu Bar* from the pop-up menu.

3 ways to move iTunes to a new device

Moving is the worst — the packing, making sure there are no mishaps along the way, putting everything back where it belongs. It's enough to make anyone steer clear — even when it comes to software. But transferring your files doesn't have to be so nerve-racking. Here are three ways to transfer your iTunes library without the moving blues.

External Drive. Moving your iTunes library from one computer to another is actually easier than it sounds. All you need is an external drive to hook up to your old computer.

Your Music folder is the default location of iTunes. Assuming all your music is being pulled from within your iTunes folder, you just need to move the folder to the external drive. This will take a while if you're a music junkie with lots of tracks.

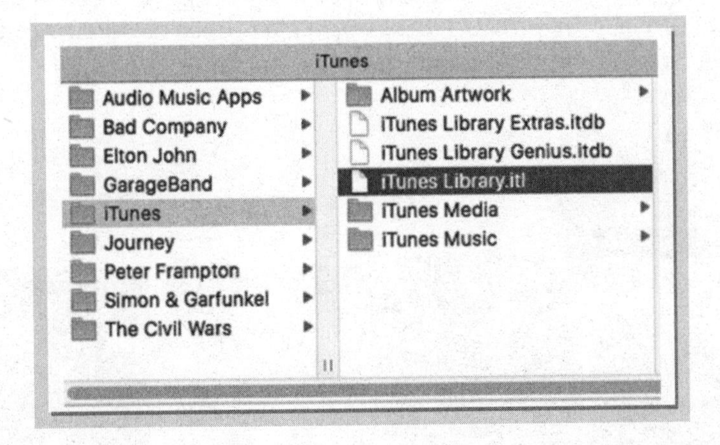

Next, hook up the external drive to your new computer. Drag the iTunes folder from the external drive to your new computer's Music folder. When you open iTunes, hold down the Option key on a Mac or Shift key in Windows. When the *Choose iTunes Library* box pops up, click *Choose Library*. Navigate to the iTunes folder you just moved to your new computer, and open the file named *iTunes Library*. Your iTunes library will be restored in all its glory.

Home Sharing. Maybe you don't want to move your iTunes library to a new computer. But you still want to listen to the music that's on your laptop from your desktop in another room. In that case, you'll love Home Sharing. This feature allows you to share and stream iTunes content with up to five computers on your home network.

Sign in to iTunes with the same Apple ID on all the devices you want to use. Go to File > Home Sharing > Turn On Home Sharing. Follow the quick steps to authorize the computer. Now you can access your music by clicking the Sharing icon (music note in a house) located in the top left corner and selecting the shared library. If you want to copy the original iTunes library to the second computer, just select the music you want to copy and click *Import*.

To use Home Sharing on your iPhone, iPad, or iPod touch, go to Settings > Music, scroll to Home Sharing, and sign in.

iTunes Match. This subscription allows you to access and download your music from all your devices and listen to your entire library no matter where you are — no external drive necessary. You can purchase iTunes Match for $24.99 a year.

When you sign in to iTunes with your Apple ID, iTunes Match turns on automatically, but if you need to turn it on manually, go to Store > iTunes Match. On an iPhone, iPad, or iPod touch, tap Settings > iTunes & App Store, and sign in. In the upper left corner, tap Settings > Music and turn on *iCloud Music Library*.

How to get iTunes music on your Android

Taking iTunes mobile is simple if you have an iPhone. But what if you have an Android? No problem. You can take your music anywhere with Google Play Music.

Sign in to your Google account on your computer. Click Google Play store > Music > My Music. To open the menu sidebar, click the three bars within a circle on the top left side of the window. Scroll down and select *Upload music*. You have to add a credit card to your account before you can import your music. But once you do, you can add 50,000 songs from iTunes to Google Play Music for free.

Next, download Google Play Music on your phone if it's not preloaded. The app works using the cloud, so you can listen to your music anywhere from any of your Android devices. Because of this, your music won't take up any room unless you choose to download your songs.

Apple earbud secrets you never knew

At first glance, they may look like regular earphones. But if you're only using Apple's EarPods to listen to music, you're missing out. The sleek-looking earbuds that come with Apple music players and iPhones have a lot packed into the cord's little remote. All you have to do is unleash its hidden powers.

First, plug your EarPods into your Apple device and press the center button once to access your music. Your earbuds will automatically find your iTunes account or other music app. Then, let the fun begin!

What you do	What happens
Press + or - volume button	Change volume
Press center button once	Play or pause music
Press center button twice quickly	Skip to next song
Press center button three times quickly	Play previous song
Press center button twice quickly and hold	Fast-forward
Press center button three times quickly and hold	Rewind
Press center button once	Answer or end call
Hold down center button for two seconds	Ignore call
Press center button once	Switch to incoming or on-hold call and put current call on hold
Hold down center button for two seconds	Switch to incoming or on-hold call and end current call
Press and hold center button	Talk to Siri
With your camera pulled up, press volume button once	Take a picture
Turn camera to video and press volume button once	Shoot a video

Index

A

B

C

D

E

F

G

H

I

J

K

L

M

N

O

P

Q

R

S

T

U

V